NumPy 1.5
Beginner's Guide

An action-packed guide for the easy-to-use, high performance, Python based free open source NumPy mathematical library using real-world examples

Ivan Idris

PUBLISHING

BIRMINGHAM - MUMBAI

NumPy 1.5
Beginner's Guide

First published: November 2011

Production Reference: 1311011

Published by Packt Publishing Ltd.
Livery Place
35 Livery Street
Birmingham B3 2PB, UK.

ISBN 978-1-84951-530-6

www.packtpub.com

Cover Image by Asher Wishkerman (wishkerman@hotmail.com)

Credits

Author

Ivan Idris

Reviewers

Lorenzo Bolla

Seth Brown

John Douglas

Finn Arup Nielsen

Ryan R. Rosario

Stefan Scherfke

Senior Acquisition Editor

Usha Iyer

Development Editor

Hyacintha D'Souza

Technical Editors

Apoorva Bolar

Aaron Rosario

Copy Editor

Brandt D'Mello

Project Coordinator

Srimoyee Ghoshal

Proofreader

Stephen Swaney

Indexer

Tejal Daruwale

Graphics

Valentina D'silva

Production Coordinator

Aparna Bhagat

Cover Work

Aparna Bhagat

About the Author

Ivan Idris has a degree in Experimental Physics and several certifications (SCJP, SCWCD and other). His graduation thesis had a strong emphasis on Applied Computer Science. After graduating, Ivan worked for several companies as Java developer, Datawarehouse developer, and Test Analyst.

More information and a blog with a few NumPy examples can be found on `ivanidris.net`

I would like to take this opportunity to thank the reviewers and the team at Packt for making this book possible.

Also, thanks goes to my teachers, professors and colleagues who taught me about science and programming.

Last, but not least; I would like to acknowledge my parents, family, and friends for their support.

About the Reviewers

Lorenzo Bolla works as Software Engineer in a successful start-up in London. His main interests are large scale web applications, numerical modelling, and functional programming.

Seth Brown is a scientist and educator with a Ph.D. in genetics/genomics from Dartmouth Medical School. He is currently employed as a bioinformatician working on deciphering novel mechanisms of human gene regulation. He has used the Python programming language in his research since 2006. He discusses his research and computational methods in his blog — drbunsen.org.

Finn Arup Nielsen is a senior researcher at the Technical University of Denmark. He has a background in machine learning and has written a PhD thesis about neuroinformatics with neuroimaging data. He has previously been using the Matlab and Perl programming languages for data processing and analysis of complex data from brain science and the Internet, but now uses more Python. Nielsen works within neuroinformatics and social media mining projects funded by the Lundbeck Foundation and The Danish Council for Strategic Research.

Ryan Rosario is a Doctoral Candidate at the University of California, Los Angeles. He works in industry as a Data Scientist and he enjoys turning large quantities of massive, messy data into gold. Ryan is heavily involved in the open-source community, particularly R, Python, Hadoop, and Machine Learning. He has also contributed code to various Python and R projects. Ryan maintains a blog dedicated to Data Science and related topics at `http://www.bytemining.com`.

Stefan Scherfke studied Computer Science with an emphasis on Environmental Computer Science at the Carl von Ossietzky University Oldenburg, Germany and received his Diplom (equiv. to M.Sc.) in 2009. Since then, he has been working in the R&D Division Energy at OFFIS—Institute for Information Technology.

In 2008, after learning various other languages (including Java, C/C++ and PHP), Stefan discovered Python and instantly fell in love with it. He has been using Python mainly to implement various simulations within the energy domain, but also to run his website and day-to-day scripting needs. He uses libraries like NumPy, SciPy, Matplotlib, SimPy, PyQt4, and Django for this. He also likes `py.test` and mock.

www.PacktPub.com

Support files, eBooks, discount offers and more

You might want to visit www.PacktPub.com for support files and downloads related to your book.

Did you know that Packt offers eBook versions of every book published, with PDF and ePub files available? You can upgrade to the eBook version at www.PacktPub.com and as a print book customer, you are entitled to a discount on the eBook copy. Get in touch with us at service@packtpub.com for more details.

At www.PacktPub.com, you can also read a collection of free technical articles, sign up for a range of free newsletters and receive exclusive discounts and offers on Packt books and eBooks.

http://PacktLib.PacktPub.com

Do you need instant solutions to your IT questions? PacktLib is Packt's online digital book library. Here, you can access, read and search across Packt's entire library of books.

Why Subscribe?

- ◆ Fully searchable across every book published by Packt
- ◆ Copy & paste, print and bookmark content
- ◆ On demand and accessible via web browser

Free Access for Packt account holders

If you have an account with Packt at www.PacktPub.com, you can use this to access PacktLib today and view nine entirely free books. Simply use your login credentials for immediate access.

To my family and friends

Table of Contents

Preface

Scientists, engineers, and quantitative data analysts face many challenges nowadays. Data scientists want to be able to do numerical analysis of large datasets with minimal programming effort. They want to write readable, efficient, and fast code, that is as close as possible to the mathematical language package they are used to. A number of accepted solutions are available in the scientific computing world.

The C, C++, and Fortran programming languages have their benefits, but they are not interactive and are considered too complex by many. The common commercial alternatives are, among others, Matlab, Maple, and Mathematica. These products provide powerful scripting languages, however, they are still more limited than any general purpose programming language. There are other open source tools similar to Matlab such as R, GNU Octave, and Scilab. Obviously, they also lack the power of a language such as Python.

Python is a popular general purpose programming language widely used by in the scientific community. You can access legacy C, Fortran, or R code easily from Python. It is object-oriented and considered more high-level than C or Fortran. Python allows you to write readable and clean code with minimal fuss. However, it lacks a Matlab equivalent out of the box. That's where NumPy comes in. This book is about NumPy and related Python libraries such as SciPy and Matplotlib.

What is NumPy?

NumPy (from **Numerical Python**) is an open source Python library for scientific computing. NumPy lets you work with arrays and matrices in a natural way. The library contains a long list of useful mathematical functions including some for linear algebra, Fourier transformation, and random number generation routines. LAPACK, a linear algebra library, is used by the NumPy linear algebra module if you have LAPACK installed on your system; otherwise NumPy provides its own implementation. LAPACK is a well known library originally written in Fortran—which Matlab relies on as well. In a sense, NumPy replaces some of the functionality of Matlab and Mathematica, allowing rapid interactive prototyping.

We will not be discussing NumPy from a developing contributor's perspective, but more from a user's perspective. NumPy is a very active project and has a lot of contributors. Maybe, one day you will be one of them!

History

NumPy is based on its predecessor, Numeric. Numeric was first released in 1995 and has a deprecated status now. Neither Numeric nor NumPy made it into the standard Python library for various reasons. However, you can install NumPy separately. More about that in the next chapter.

In 2001, a number of people inspired by Numeric created SciPy—an open source Python scientific computing library that provides functionality similar to that of Matlab, Maple, and Mathematica. Around this time, people were growing increasingly unhappy with Numeric. Numarray was created as alternative for Numeric. Numarray is currently also deprecated. Numarray was better in some areas than Numeric, but worked very differently. For that reason, SciPy kept on depending on the Numeric philosophy and the Numeric array object. As is customary with new "latest and greatest" software, the arrival of Numarray led to the development of an entire whole ecosystem around it with a range of useful tools. Unfortunately, the SciPy community could not enjoy the benefits of this development. It is quite possible that some Pythonista has decided to neither choose neither one nor the other camp.

In 2005, Travis Oliphant, an early contributor to SciPy, decided to do something about this situation. He tried to integrate some of the Numarray features into Numeric. A complete rewrite took place that culminated into the release of NumPy 1.0 in 2006. At this time, NumPy has all of the features of Numeric and Numarray and more. Upgrade tools are available to facilitate the upgrade from Numeric and Numarray. The upgrade is recommended since Numeric and Numarray are not actively supported any more.

Originally the NumPy code was part of SciPy. It was later separated and is now used by SciPy for array and matrix processing.

Why use NumPy?

NumPy code is much cleaner than "straight" Python code that tries to accomplish the same task. There are fewer loops required because operations work directly on arrays and matrices. The many convenience and mathematical functions make life easier as well. The underlying algorithms have stood the test of time and have been designed with high performance in mind.

NumPy's arrays are stored more efficiently than an equivalent data structure in base Python such as a list of lists. Array I/O is significantly faster too. The performance improvement scales with the number of elements of an array. It really pays off to use NumPy for large arrays. Files as large as several terabytes can be memory-mapped to arrays leading to optimal reading and writing of data. The drawback of NumPy arrays is that they are more specialized than plain lists. Outside of the context of numerical computations, NumPy arrays are less useful. The technical details of NumPy arrays will be discussed in later chapters.

Large portions of NumPy are written in C. That makes NumPy faster than pure Python code. A NumPy C API exists as well. It allows further extension of the functionality with the help of the C language of NumPy. The C API falls outside the scope of the book. Finally, since NumPy is open source, you get all the added advantages. The price is the lowest possible—free as in 'beer'. You don't have to worry about licenses every time somebody joins your team or you need an upgrade of the software. The source code is available to everyone. This, of course, is beneficial to the code quality.

Limitations of NumPy

There is one important thing to know if you are planning to create Google App Engine applications. NumPy is not supported within the Google App Engine sandbox. NumPy is deemed "unsafe" partly because it is written in C.

If you are a Java programmer, you may be interested in Jython, the Java implementation of Python. In that case, I have bad news for you. Unfortunately, Jython runs on the Java Virtual Machine and cannot access NumPy because NumPy's modules are mostly written in C. You could say that Jython and Python are from two totally different worlds, although they do implement the same specification.

The stable release of NumPy, at the time of writing, supported Python 2.4 to 2.6.x, and now also supports Python 3.

What this book covers

Chapter 1, NumPy Quick Start, will guide you through the steps needed to install NumPy on your system and create a basic NumPy application.

Chapter 2, Beginning with NumPy Fundamentals, introduces you to NumPy arrays and fundamentals.

Chapter 3, Get into Terms with Commonly Used Functions, will teach you about the most commonly used NumPy functions—the basic mathematical and statistical functions.

Chapter 4, Convenience Functions for Your Convenience, will teach you about functions that make working with NumPy easier. This includes functions that select certain parts of your arrays, for instance based on a Boolean condition. You will also learn about polynomials and manipulating the shape of NumPy objects.

Chapter 5, Working with Matrices and ufuncs, covers matrices and universal functions. Matrices are well known in mathematics and have their representation in NumPy as well. Universal functions (ufuncs) work on arrays element-by-element or on scalars. ufuncs expect a set of scalars as input and produce a set of scalars as output.

Chapter 6, Move Further with NumPy Modules, discusses how universal functions can typically be mapped to mathematical counterparts such as add, subtract, divide, multiply, and so on. NumPy has a number of basic modules that will be discussed in this chapter.

Chapter 7, Peeking into Special Routines, describes some of the more specialized NumPy functions. As NumPy users, we sometimes find ourselves having special needs. Fortunately, NumPy provides for most of our needs.

Chapter 8, Assured Quality with Testing, will teach you how to write NumPy unit tests.

Chapter 9, Plotting with Matplotlib, discusses how NumPy on its own cannot be used to create graphs and plots. This chapter covers (in-depth) Matplotlib, a very useful Python plotting library. Matplotlib integrates nicely with NumPy and has plotting capabilities comparable to Matlab.

Chapter 10, When NumPy is Not Enough: SciPy and Beyond, discuss how SciPy and NumPy are historically related. This chapter goes into more detail about SciPy. SciPy, as mentioned in the *History* section, is a high level Python scientific computing framework built on top of NumPy. It can be used in conjunction with NumPy.

What you need for this book

To try out the code samples in this book, you will need a recent build of NumPy. This means that you will need to have one of the Python versions supported by NumPy as well. Some code samples make use of Matplotlib for illustration purposes. Matplotlib is not strictly required to follow the examples, but it is recommended that you install it too. The last chapter is about SciPy and has one example involving SciKits.

Here is a list of software used to develop and test the code examples:

- ◆ Python 2.6
- ◆ NumPy 2.0.0.dev20100915
- ◆ SciPy 0.9.0.dev20100915

- Matplotlib 1.0.0
- Ipython 0.10

Needless to say, you don't need to have exactly this software and these versions on your computer. Python and NumPy is the absolute minimum you will need.

Who this book is for

This book is for you the scientist, engineer, programmer, or analyst looking for a high quality open source mathematical library. Knowledge of Python is assumed. Also, some affinity or at least interest in mathematics and statistics is required.

Conventions

In this book, you will find several headings appearing frequently.

To give clear instructions of how to complete a procedure or task, we use:

Time for action – heading

1. Action 1
2. Action 2
3. Action 3

Instructions often need some extra explanation so that they make sense, so they are followed with:

What just happened?

This heading explains the working of tasks or instructions that you have just completed.

You will also find some other learning aids in the book, including:

Pop quiz – heading

These are short multiple choice questions intended to help you test your own understanding.

Have a go hero – heading

These set practical challenges and give you ideas for experimenting with what you have learned.

You will also find a number of styles of text that distinguish between different kinds of information. Here are some examples of these styles, and an explanation of their meaning.

Code words in text are shown as follows: "We can include other contexts through the use of the `include` directive."

A block of code is set as follows:

```
[def pythonsum(n):
  a = range(n)
  b = range(n)
  c = []
  for i in range(len(a)):
    a[i] = i ** 2
    b[i] = i ** 3
    c.append(a[i] + b[i])
    return c
```

When we wish to draw your attention to a particular part of a code block, the relevant lines or items are set in bold:

```
[def pythonsum(n):
  a = range(n)
  b = range(n)
  c = []
  for i in range(len(a)):
    a[i] = i ** 2
    b[i] = i ** 3
    c.append(a[i] + b[i])
    return c
```

Any command-line input or output is written as follows:

```
sudo apt-get install python
```

New terms and **important words** are shown in bold. Words that you see on the screen, in menus or dialog boxes for example, appear in the text like this: "clicking the **Next** button moves you to the next screen".

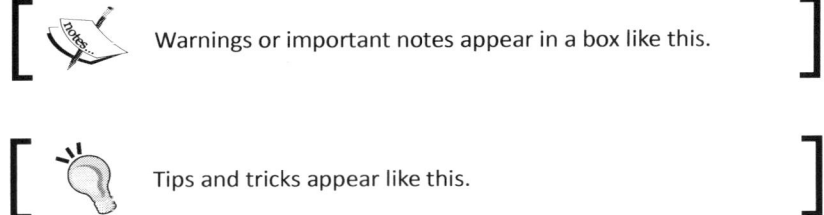

Warnings or important notes appear in a box like this.

Tips and tricks appear like this.

Reader feedback

Feedback from our readers is always welcome. Let us know what you think about this book—what you liked or may have disliked. Reader feedback is important for us to develop titles that you really get the most out of.

To send us general feedback, simply send an e-mail to feedback@packtpub.com, and mention the book title via the subject of your message.

If there is a book that you need and would like to see us publish, please send us a note in the **SUGGEST A TITLE** form on www.packtpub.com or e-mail suggest@packtpub.com.

If there is a topic that you have expertise in and you are interested in either writing or contributing to a book, see our author guide on www.packtpub.com/authors.

Customer support

Now that you are the proud owner of a Packt book, we have a number of things to help you to get the most from your purchase.

Downloading the example code

You can download the example code files for all Packt books you have purchased from your account at http://www.PacktPub.com. If you purchased this book elsewhere, you can visit http://www.PacktPub.com/support and register to have the files e-mailed directly to you.

Errata

Although we have taken every care to ensure the accuracy of our content, mistakes do happen. If you find a mistake in one of our books—maybe a mistake in the text or the code—we would be grateful if you would report this to us. By doing so, you can save other readers from frustration and help us improve subsequent versions of this book. If you find any errata, please report them by visiting http://www.packtpub.com/support, selecting your book, clicking on the **errata submission form** link, and entering the details of your errata. Once your errata are verified, your submission will be accepted and the errata will be uploaded on our website, or added to any list of existing errata under the Errata section of that title. Any existing errata can be viewed by selecting your title from http://www.packtpub.com/support.

Piracy

Piracy of copyright material on the Internet is an ongoing problem across all media. At Packt, we take the protection of our copyright and licenses very seriously. If you come across any illegal copies of our works, in any form, on the Internet, please provide us with the location address or website name immediately so that we can pursue a remedy.

Please contact us at copyright@packtpub.com with a link to the suspected pirated material.

We appreciate your help in protecting our authors, and our ability to bring you valuable content.

Questions

You can contact us at questions@packtpub.com if you are having a problem with any aspect of the book, and we will do our best to address it.

1
NumPy Quick Start

Let's get started. We will install NumPy on different operating systems and have a look at some simple code that uses NumPy. The IPython interactive shell is introduced briefly. As mentioned in the preface, SciPy is closely related to NumPy, so you will see the SciPy name appearing here and there. At the end of this chapter, you will find pointers on how to find additional information online if you get stuck or are uncertain about the best way to solve problems.

In this chapter, we shall:

- Install Python and NumPy on Windows
- Install Python and NumPy on Linux
- Install Python and NumPy on Macintosh
- Write simple NumPy code
- Get to know IPython
- Browse online documentation and resources

Python

NumPy is based on Python, so it is required to have Python installed. On some operating systems, Python is already installed. You, however, need to check whether the Python version corresponds with the NumPy version you want to install.

Time for action – installing Python on different operating systems

NumPy has binary installers for Windows, various Linux distributions and Mac OS X. There is also a source distribution, if you prefer that. You need to have Python 2.4.x or above installed on your system. We will go through the various steps required to install Python on the following operating systems:

1. **Debian and Ubuntu**: Python might already be installed on Debian and Ubuntu but the development headers are usually not. On Debian and Ubuntu install python and python-dev with the following commands:

   ```
   sudo apt-get install python
   sudo apt-get install python-dev
   ```

2. **Windows**: The Windows Python installer can be found at `www.python.org/download`. On this website, we can also find installers for Mac OS X and source tarballs for Linux, Unix, and Mac OS X.

3. **Mac**: Python comes pre-installed on Mac OS X. We can also get Python through MacPorts, Fink, or similar projects.

   We can install, for instance, the Python 2.6 port by running the following command:

   ```
   sudo port install python26
   ```

   LAPACK does not need to be present but, if it is, NumPy will detect it and use it during the installation phase. It is recommended to install LAPACK for serious numerical analysis.

What just happened?

We installed Python on Debian, Ubuntu, Windows, and the Mac.

> **Downloading the example code**
>
> You can download the example code files for all Packt books you have purchased from your account at `http://www.PacktPub.com`. If you purchased this book elsewhere, you can visit `http://www.PacktPub.com/support` and register to have the files e-mailed directly to you.

Windows

Installing NumPy on Windows is straightforward. You only need to download an installer, and a wizard will guide you through the installation steps.

Time for action – installing NumPy on Windows

Installing NumPy on Windows is necessary but, fortunately, a straightforward task. The actions we will take are as follows:

1. **Download the NumPy installer**: Download a NumPy installer for Windows from the SourceForge website `http://sourceforge.net/projects/numpy/files/`

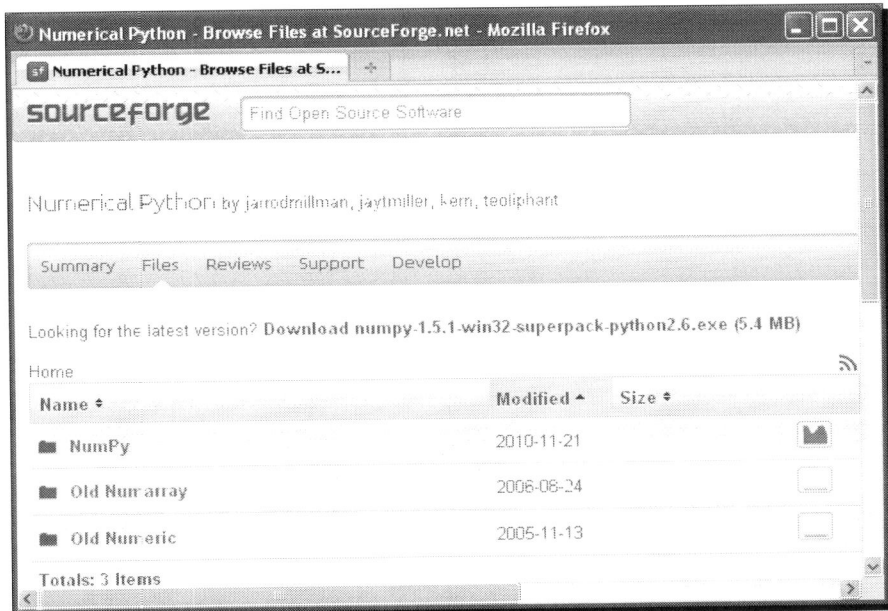

Choose the appropriate version. In this example, we chose `numpy-1.5.1-win32-superpack-python2.6.exe`.

2. **Open the installer**: Open the EXE installer by double clicking on it.

3. **NumPy features**: Now, we see a description of NumPy and its features. Click **Next**.

4. **Install Python**: If you have Python installed, it should automatically be detected. If it is not detected, maybe your path settings are wrong. At the end of this chapter, resources are listed in case you have problems with installing NumPy:

5. **Finish the installation**: In this example, Python 2.6 was found. Click **Next** if Python is found; otherwise, click **Cancel** and install Python (NumPy cannot be installed without Python). Click **Next**. This is the point of no return. Well, kind of, but it is best to make sure that you are installing to the proper directory and so on and so forth. Now the real installation starts. This may take a while:

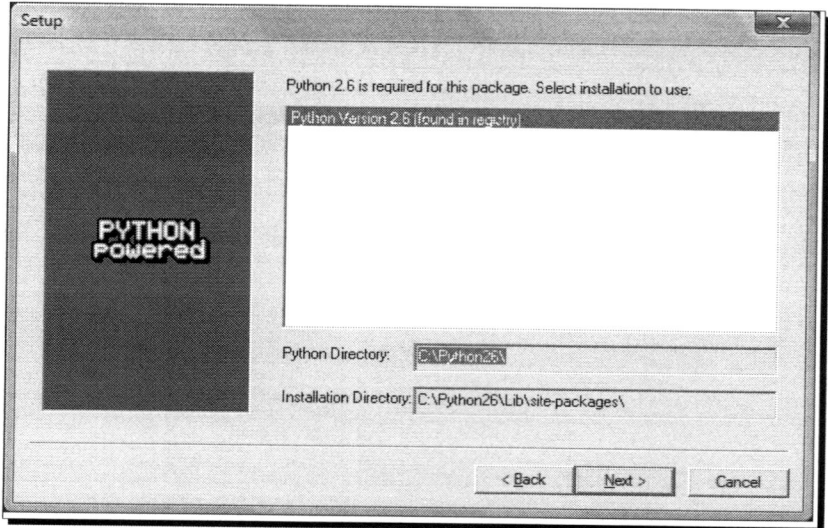

What just happened?

We installed NumPy on Windows.

Linux

Installing NumPy on Linux depends on the distribution you have. We will discuss how you would install NumPy from the command line, although you could probably use graphical installers; it depends on your **distribution (distro)**.

Time for action – installing NumPy on Linux

Most Linux distributions have NumPy packages. We will go through the necessary steps for some of the popular Linux distros:

1. **Installing NumPy on Red Hat**: Run the following instructions from the command line:

   ```
   yum install python-numpy
   ```

2. **Installing NumPy on Mandriva**: To install NumPy on Mandriva, run the following command line instruction:

   ```
   urpmi python-numpy
   ```

3. **Installing NumPy on Gentoo**: To install NumPy on Gentoo run the following command line instruction:

   ```
   sudo emerge numpy
   ```

4. **Installing NumPy on Debian and Ubuntu**: On Debian or Ubuntu, we need to type the following:

   ```
   sudo apt-get install python-numpy
   ```

The following table gives an overview of the Linux distributions and corresponding NumPy package names.

Linux distribution	Package name
Arch Linux	python-numpy
Debian	python-numpy
Fedora	numpy
Gentoo	dev-python/numpy
OpenSUSE	python-numpy, python-numpy-devel
Slackware	numpy

What just happened?

We installed NumPy on various Linux distributions.

Mac OS X

You can install NumPy on the Mac with a graphical installer or from the command-line from a port manager such as MacPorts or Fink, depending on your preference.

Time for action – installing NumPy on Mac OS X with a GUI installer

We will install NumPy with a GUI installer.

1. **Download the GUI installer**: We can get a NumPy installer from the SourceForge website `http://sourceforge.net/projects/numpy/files/`. Download the appropriate DMG file. Usually the latest one is the best:

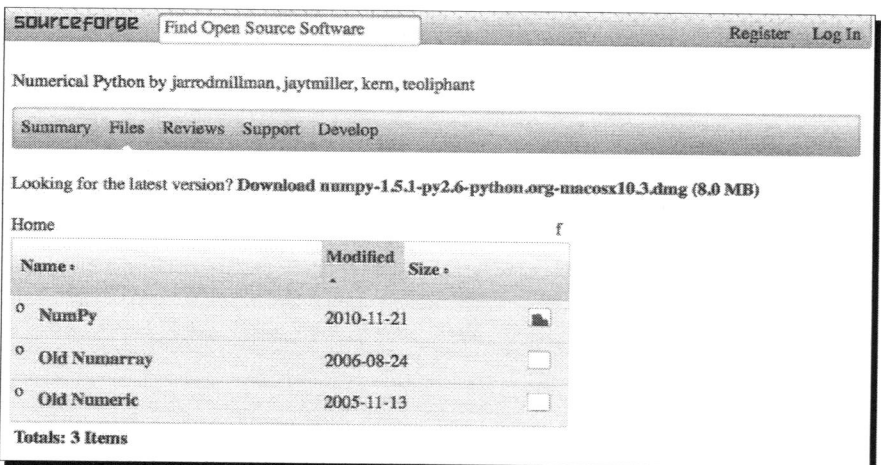

2. **Open the DMG file**: Open the DMG file (in this example, `numpy-1.5.1-py2.6-python.org-macosx10.3.dmg`):

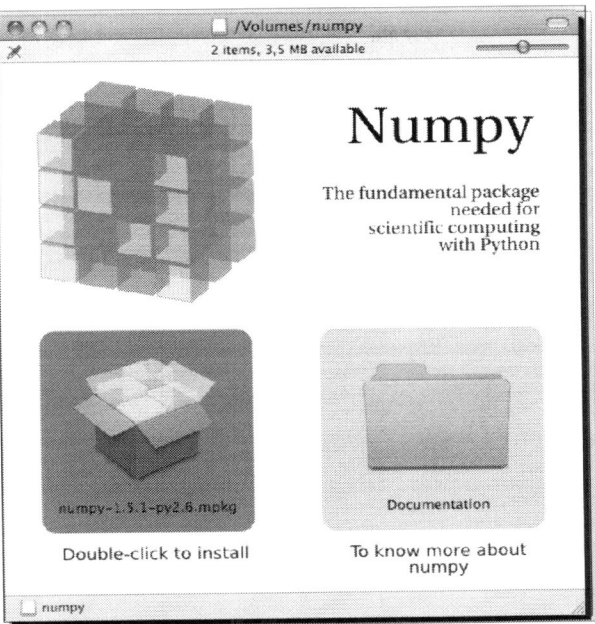

☐ Double-click on the icon of the opened box, the one having a subscript that ends with **.mpkg**. We will be presented with the welcome screen of the installer.

☐ Click on the **Continue** button to go to the **Read Me** screen, where we will be presented with a short description of NumPy:

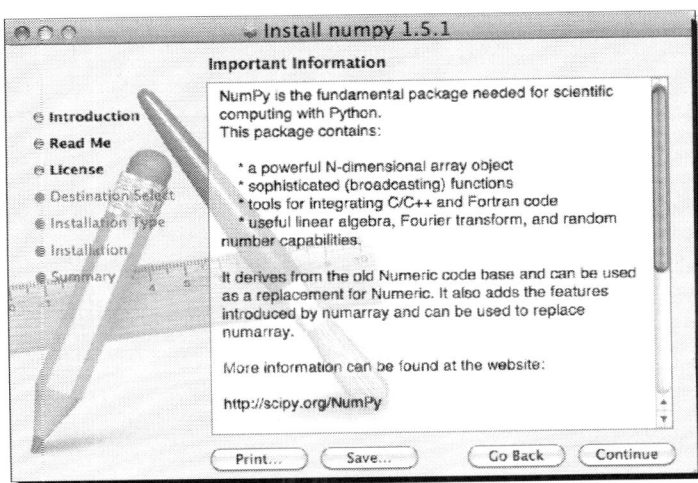

☐ Continue to the License screen.

3. **Accept the license**: Read the license, click **Continue** and then the **Accept** button, when prompted to accept the license. Continue through the next screens and click **Finish** at the end.

What just happened?

We installed NumPy on Mac OS X with a GUI installer.

Time for action – installing NumPy with MacPorts or Fink

Alternatively we can install NumPy through the MacPorts route. It is shown as follows:

1. Installing with MacPorts: Type the following command:

   ```
   sudo port install py-numpy
   ```

2. Installing with Fink: Fink also has packages for NumPy—scipy-core-py24, scipy-core-py25, and scipy-core-py26. We can install the one for Python 2.6 with the following package:

   ```
   fink install scipy-core-py26
   ```

What just happened?

We installed NumPy on Mac OS X with MacPorts and Fink.

Building from source

We can retrieve the source code for NumPy with git. This is shown as follows:

```
git clone git://github.com/numpy/numpy.git numpy
```

Install /usr/local with the following command:

```
python setup.py build
sudo python setup.py install --prefix=/usr/local
```

To build, we need a C compiler such as GCC and the Python header files in the python-dev or python-devel package.

Vectors

NumPy arrays are more efficient than Python lists, when it comes to numerical operations. NumPy code requires less explicit loops than equivalent Python code.

Time for action – adding vectors

Imagine that we want to add two vectors called a and b. Vector a holds the squares of integers 0 to n, for instance, if n = 3, then a = (0, 1, 4). Vector b holds the cubes of integers 0 to n, so if n = 3, then b = (0, 1, 8). How would you do that using plain Python? After we come up with a solution, we will compare it with the NumPy equivalent.

1. **Adding vectors using pure Python**: The following function solves the vector addition problem using pure Python without NumPy:

```
def pythonsum(n):
    a = range(n)
    b = range(n)
    c = []

    for i in range(len(a)):
        a[i] = i ** 2
        b[i] = i ** 3
        c.append(a[i] + b[i])

    return c
```

2. **Adding vectors using NumPy**: Following is a function that achieves the same with NumPy.

```
def numpysum(n):
    a = numpy.arange(n) ** 2
    b = numpy.arange(n) ** 3
    c = a + b
    return c
```

Notice that numpysum() does not need a for loop. Also, we used the arange function from NumPy that creates a NumPy array for us with integers 0 to n. The arange function was imported; that is why it is prefixed with numpy.

Now comes the fun part. Remember that it is mentioned in the preface that NumPy is faster when it comes to array operations. How much faster is Numpy, though? The following program will show us by measuring the elapsed time in microseconds, for the numpysum and pythonsum functions. It also prints the last two elements of the vector sum. Let's check that we get the same answers by using Python and NumPy:

```
import sys
from datetime import datetime
import numpy

def numpysum(n):
    a = numpy.arange(n) ** 2
```

```
        b = numpy.arange(n) ** 3
        c = a + b
        return c

    def pythonsum(n):
        a = range(n)
        b = range(n)
        c = []

        for i in range(len(a)):
            a[i] = i ** 2
            b[i] = i ** 3
            c.append(a[i] + b[i])

        return c

    size = int(sys.argv[1])
    start = datetime.now()
    c = pythonsum(size)
    delta = datetime.now() - start
    print "The last 2 elements of the sum", c[-2:]
    print "PythonSum elapsed time in microseconds", delta.microseconds
    start = datetime.now()
    c = numpysum(size)
    delta = datetime.now() - start
    print "The last 2 elements of the sum", c[-2:]
    print "NumPySum elapsed time in microseconds", delta.microseconds
```

The output of the program for 1000, 2000, and 3000 vector elements is as follows:

```
$ python vectorsum.py 1000
The last 2 elements of the sum [995007996, 998001000]
PythonSum elapsed time in microseconds 707
The last 2 elements of the sum [995007996 998001000]
NumPySum elapsed time in microseconds 171


$ python vectorsum.py 2000
The last 2 elements of the sum [7980015996, 7992002000]
PythonSum elapsed time in microseconds 1420
The last 2 elements of the sum [7980015996 7992002000]
NumPySum elapsed time in microseconds 168
```

```
$ python vectorsum.py 4000
The last 2 elements of the sum [63920031996, 63968004000]
PythonSum elapsed time in microseconds 2829
The last 2 elements of the sum [63920031996 63968004000]
NumPySum elapsed time in microseconds 274
```

What just happened?

Clearly, NumPy is much faster than the equivalent normal Python code. One thing is certain; we get the same results whether we are using NumPy or not. However, the result that is printed differs in representation. Notice that the result from the numpysum function does not have any commas. How come? Obviously we are not dealing with a Python list but with a NumPy array. It was mentioned in the preface that NumPy arrays are specialized data structures for numerical data. We will learn more about NumPy arrays in the next chapter.

Pop Quiz – functioning of arange function

1. What does arange(5) do?

   ❑ Creates a Python list of 5 elements with values 1 to 5.

   ❑ Creates a Python list of 5 elements with values 0 to 4.

   ❑ Creates a NumPy array with values 1 to 5.

   ❑ Creates a NumPy array with values 0 to 4. ✓

   ❑ None of the above.

Have a go hero – continue the analysis

The program we used here to compare the speed of NumPy and regular Python is not very scientific. We should at least repeat each measurement a couple of times. It would be nice to be able to calculate some statistics such as average times, and so on. Also, you might want to show plots of the measurements to friends and colleagues.

 Hints to help you can be found throughout this book and in the online documentation and resources listed at the end of this chapter. NumPy has, by the way, statistical functions that can calculate averages for you. I recommend using matplotlib to produce plots.

IPython—an interactive shell

Scientists and engineers are used to experimenting. IPython was created by scientists with experimentation in mind. The interactive environment that IPython provides is viewed by many as a direct answer to Matlab, Mathematica, and Maple. You can find more information, including installation instructions, at: http://ipython.org/

IPython is free, open source, and available for Linux, Unix, Mac OS X, and Windows. The IPython authors only request that you cite IPython in scientific work where IPython was used. Here is the list of features of IPython:

◆ Tab completion

◆ History mechanism

◆ Inline editing

◆ Ability to call external Python scripts with %run

◆ Access to system commands

◆ Pylab switch

◆ Access to Python debugger and profiler

The Pylab switch imports all the Scipy, NumPy, and Matplotlib packages. Without this switch, we would have to import every package we need ourselves.

All we need to do is enter the following instruction on the command line:

```
$ ipython -pylab
Python 2.6.1 (r261:67515, Jun 24 2010, 21:47:49)
Type "copyright", "credits" or "license" for more information.
IPython 0.10 -- An enhanced Interactive Python.
?         -> Introduction and overview of IPython's features.
%quickref -> Quick reference.
help      -> Python's own help system.
object?   -> Details about 'object'. ?object also works, ?? prints more.
  Welcome to pylab, a matplotlib-based Python environment.
  For more information, type 'help(pylab)'.
In [1]: quit()
```

quit() or *Ctrl + D* quits the IPython shell. We might want to be able to go back to our experiments. In IPython, it is easy to save a session for later.

```
In [1]: %logstart
Activating auto-logging. Current session state plus future input saved.
```

```
Filename        : ipython_log.py
Mode            : rotate
Output logging  : False
Raw input log   : False
Timestamping    : False
State           : active
```

Let's say we have the vector addition program that we made in the current directory. We can run the script as follows:

```
In [1]: ls
README          vectorsum.py
In [2]: %run -i vectorsum.py 1000
```

As you probably remember, 1000 specifies the number of elements in a vector. The -d switch of `%run` starts an `ipdb` debugger with '*c*' the script is started. 'n' steps through the code. Typing `quit` at the `ipdb` prompt exits the debugger.

```
In [2]: %run -d vectorsum.py 1000
*** Blank or comment
*** Blank or comment
Breakpoint 1 at: /Users/ivanidris/Documents/numpyBeginnersGuide/book/
ch1code/vectorsum.py:3
```

 Enter *c* at the `ipdb>` prompt to start your script.

```
><string>(1)<module>()
ipdb> c
> /Users/ivanidris/Documents/numpyBeginnersGuide/book/ch1code/vectorsum.
py(3)<module>()
        2
1---> 3 import sys
        4 from datetime import datetime
ipdb> n
>
/Users/ivanidris/Documents/numpyBeginnersGuide/book/ch1code/vectorsum.
py(4)<module>()
1       3 import sys
----> 4 from datetime import datetime
```

```
      5 import numpy
ipdb> n
> /Users/ivanidris/Documents/numpyBeginnersGuide/book/ch1code/vectorsum.
py(5)<module>()
      4 from datetime import datetime
----> 5 import numpy
      6
ipdb> quit
```

We can also profile our script by passing the `-p` option to `%run`.

```
In [4]: %run -p vectorsum.py 1000
        1058 function calls (1054 primitive calls) in 0.002 CPU seconds
   Ordered by: internal time

ncallstottimepercallcumtimepercallfilename:lineno(function)
1 0.001    0.001    0.001      0.001 vectorsum.py:28(pythonsum)
1 0.001    0.001    0.002      0.002 {execfile}
1000 0.000    0.0000.0000.000 {method 'append' of 'list' objects}
1 0.000    0.000    0.002      0.002 vectorsum.py:3(<module>)
1 0.000    0.0000.0000.000 vectorsum.py:21(numpysum)
3    0.000    0.0000.0000.000 {range}
1    0.000    0.0000.0000.000 arrayprint.py:175(_array2string)
3/1    0.000     0.0000.0000.000 arrayprint.py:246(array2string)
2    0.000    0.0000.0000.000 {method 'reduce' of 'numpy.ufunc' objects}
4    0.000    0.0000.0000.000 {built-in method now}
2    0.000    0.0000.0000.000 arrayprint.py:486(_formatInteger)
2    0.000    0.0000.0000.000 {numpy.core.multiarray.arange}
1    0.000    0.0000.0000.000 arrayprint.py:320(_formatArray)
3/1    0.000     0.0000.0000.000 numeric.py:1390(array_str)
1    0.000    0.0000.0000.000 numeric.py:216(asarray)
2    0.000    0.0000.0000.000 arrayprint.py:312(_extendLine)
1    0.000    0.0000.0000.000 fromnumeric.py:1043(ravel)
2    0.000    0.0000.0000.000 arrayprint.py:208(<lambda>)
1    0.000    0.000    0.002      0.002<string>:1(<module>)
11    0.000    0.0000.0000.000 {len}
2    0.000    0.0000.0000.000 {isinstance}
1    0.000    0.0000.0000.000 {reduce}
```

```
1    0.000    0.0000.0000.000 {method 'ravel' of 'numpy.ndarray' objects}
4    0.000    0.0000.0000.000 {method 'rstrip' of 'str' objects}
3    0.000    0.0000.0000.000 {issubclass}
2    0.000    0.0000.0000.000 {method 'item' of 'numpy.ndarray' objects}
1    0.000    0.0000.0000.000 {max}
1    0.000    0.0000.0000.000 {method 'disable' of '_lsprof.Profiler'
objects}
```

This gives us a bit more insight in the workings of our program. In addition, we can now identify performance bottlenecks. The %hist command shows the commands history.

```
In [2]: a=2+2
In [3]: a
Out[3]: 4
In [4]: %hist
1: _ip.magic("hist ")
2: a=2+2
3: a
```

I hope you agree that IPython is a really useful tool!

Online resources and help

When we are in IPython's pylab mode, we can open manual pages for NumPy functions with the help command. It is not necessary to know the name of a function. We can type a few characters and then let tab completion do its work. Let's, for instance, browse the available information for the arange function.

```
In [2]: help ar<Tab>

arangearccosarccosharcsinarcsinh
arctan arctan2 arctanhargmaxargmin
argsortargwhere around array2string array_equal
array_equivarray_reprarray_splitarray_str arrow
array
In [2]: help arange
```

Another option is to put a question mark behind the function name.

```
In [3]: arange?
```

The main documentation website for NumPy and SciPy is at `http://docs.scipy.org/doc/`. Through this webpage, we can browse the NumPy reference at `http://docs.scipy.org/doc/numpy/reference/` and the user guide as well as several tutorials.

NumPy has a wiki with lots of documentation at `http://docs.scipy.org/numpy/Front%20Page/`.

The NumPy and SciPy forum can be found at `http://ask.scipy.org/en`.

The popular Stack Overflow software development forum has hundreds of questions tagged "numpy". To view them, go to `http://stackoverflow.com/questions/tagged/numpy`.

If you are really stuck with a problem or you want to be kept informed of NumPy development, you can subscribe to the NumPy discussion mailing list. The e-mail address is `numpy-discussion@scipy.org`. The number of e-mails per day is not too high and there is almost no spam to speak of. Most importantly, developers actively involved with NumPy also answer questions asked on the discussion group. The complete list can be found at `http://www.scipy.org/Mailing_Lists`.

For IRC users, there is an IRC channel on `irc.freenode.net`. The channel is called **#scipy**, but you can also ask NumPy questions since SciPy users also have knowledge of NumPy, as SciPy is based on NumPy. There are at least 50 members on the scipy channel at all times.

Summary

In this chapter, we installed NumPy. We got a vector addition program working and convinced ourselves that NumPy has superior performance. We were introduced to the IPython interactive shell. In addition, we explored the available NumPy documentation and online resources.

In the next chapter, we will take a look under the hood and explore some fundamental concepts including arrays and data types.

2
Beginning with NumPy Fundamentals

After installing NumPy and getting some code to work, it's time to cover NumPy basics.

The topics we shall cover in this chapter are:

◆ Data types

◆ Array types

◆ Type conversions

◆ Array creation

◆ Indexing

◆ Slicing

◆ Shape manipulation

Before we start, let me make a few remarks about the code examples in this chapter. The code snippets in this chapter show input and output from several IPython sessions. Recall that IPython was introduced in the previous chapter as the interactive Python shell of choice for scientific computing. The advantages of IPython are pylab switch of many scientific computing Python packages, including NumPy, and the fact that it is not necessary to explicitly call the `print` function to display variable values. However, the source code delivered alongside the book is regular Python code that uses imports and print statements.

NumPy array object

NumPy has a multi-dimensional array object called ndarray.It consists of two parts:

1. The actual data
2. Some metadata describing the data

The majority of array operations leave the raw data untouched. The only aspect that changes is the metadata.

We have already learned, in the previous chapter, how to create an array using the arange function. Actually, we created a one-dimensional array that contained a set of numbers. ndarray can have more than one dimension.

The NumPy array is homogeneous—the items in the array have to be of the same type. The advantage is that, if we know that the items in the array are of the same type, then it is easy to determine the storage size required for the array.

NumPy arrays are indexed just like in Python, starting from 0. Data types are represented by special objects. These objects will be discussed comprehensively in this chapter.

We will create an array with the arange function again. Here's how to get the data type of an array:

```
In: a = arange(5)
In: a.dtype
Out: dtype('int64')
```

The data type of array a is int64 (at least on my machine), but you may get int32 as output if you are using 32-bit Python. In both cases, we are dealing with integers (64-bit or 32-bit). Besides the data type of an array, it is important to know its shape. The following diagram will give us a better understanding of a NumPy array object:

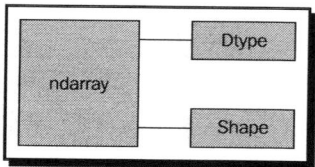

The example in *Chapter 1, NumPy Quick Start*, demonstrated how to create a vector (actually, a one-dimensional NumPy array). A vector is commonly used in mathematics but, most of the time, we need higher-dimensional objects. Let's determine the shape of the vector we created a few minutes ago:

```
In [4]: a
Out[4]: array([0, 1, 2, 3, 4])
In: a.shape
Out: (5,)
```

As you can see, the vector has five elements with values ranging from 0 to 4. The shape attribute of the array is a tuple, in this case a tuple of 1 element, which contains the length in each dimension.

Time for action – creating a multidimensional array

Now that we know how to create a vector, we are ready to create a multidimensional NumPy array. After we create the matrix, we would again want to display its shape and data type.

1. Create a multidimensional array.

2. Show the array shape and data type:

```
In: m = array([arange(2), arange(2)])
In: m
Out:
array([[0, 1],
       [0, 1]])
In: m.shape
Out: (2, 2)
```

What just happened?

We created a 2-by-2 array with the arange function we have come to trust and love. Without any warning, the array function appeared on the stage.

The array function creates an array from an object that you give to it. The object needs to be array-like, for instance, a Python list. In the preceding example, we passed in a list of arrays. The object is the only required argument of the array function. NumPy functions tend to have a lot of optional arguments with predefined defaults.

Pop quiz – the shape of ndarray

1. How is the shape of an ndarray stored?

    a. It is stored in a comma-separated string.

    b. It is stored in a list.

    c. It is stored in a tuple.

Have a go hero – create a 3-by-3 matrix

It shouldn't be too hard now to create a 3-by-3 matrix. Give it a go and check whether the array shape is as expected.

Selecting elements

From time to time, we will want to select a particular element of an array. We will take a look at how to do this, but first, let's create a 2-by-2 matrix again:

```
In: a = array([[1,2],[3,4]])
In: a
Out:
array([[1, 2],
       [3, 4]])
```

The matrix was created this time by passing the array function a list of lists. We will now select each item of the matrix one-by-one. Remember, the indices are numbered starting from 0.

```
In: a[0,0]
Out: 1
In: a[0,1]
Out: 2
In: a[1,0]
Out: 3
In: a[1,1]
Out: 4
```

As you can see, selecting elements of the array is pretty simple. For the array a, we just use the notation a[m,n] , where m and n are the indices of the item in the array.

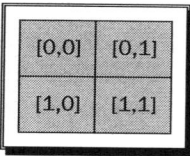

NumPy numerical types

Python has an integer type, a float type, and a complex type, however, this is not enough for scientific computing and, for this reason, NumPy has a lot more data types. In practice, we need even more types with varying precision and, therefore, different memory size of the type. The majority of the NumPy numerical types end with a number. This number indicates the number of bits associated with the type. The following table (adapted from the NumPy user guide) gives an overview of NumPy numerical types:

Type	Description
bool	Boolean (True or False) stored as a bit
inti	Platform integer (normally either int32 or int64)
int8	Byte (-128 to 127)
int16	Integer (-32768 to 32767)
int32	Integer (-2 ** 31 to 2 ** 31 -1)
int64	Integer (-2 ** 63 to 2 ** 63 -1)
uint8	Unsigned integer (0 to 255)
uint16	Unsigned integer (0 to 65535)
uint32	Unsigned integer (0 to 2 ** 32 - 1)
uint64	Unsigned integer (0 to 2 ** 64 - 1)
float16	Half precision float: sign bit, 5 bits exponent, 10 bits mantissa
float32	Single precision float: sign bit, 8 bits exponent, 23 bits mantissa
float64 or float	Double precision float: sign bit, 11 bits exponent, 52 bits mantissa
complex64	Complex number, represented by two 32-bit floats (real and imaginary components)
complex128 or complex	Complex number, represented by two 64-bit floats (real and imaginary components)

For each data type, there exists a corresponding conversion function:

```
In: float64(42)
Out: 42.0
In: int8(42.0)
Out: 42
In: bool(42)
Out: True
In: bool(42.0)
Out: True
In: float(True)
Out: 1.0
```

Many functions have a data type argument, which is often optional:

```
In: arange(7, dtype=uint16)
Out: array([0, 1, 2, 3, 4, 5, 6], dtype=uint16)
```

It is important to know that you are not allowed to convert a complex number into an integer. Trying to do that triggers a `TypeError`. This is shown as follows:

```
In: int(42.0 + 1.j)
------------------------------------------------------------------
---
TypeError                                   Traceback (most recent call
last)
TypeError: can't convert complex to int; use int(abs(z))
```

The same goes for conversion of a complex number into a float. By the way, the `.j` part is the imaginary coefficient of the complex number. See the following code:

```
In: float(42.0 + 1.j)
------------------------------------------------------------------
---
TypeError                                   Traceback (most recent call
last)
TypeError: can't convert complex to float; use abs(z)
```

Data type objects

Data type objects are instances of the `numpy.dtype` class. Once again, arrays have a data type. To be precise, every element in a NumPy array has the same data type. The data type object can tell you the size of the data in bytes. The size in bytes is given by the `itemsize` attribute of the `dtype` class:

```
In: a.dtype.itemsize
Out: 8
```

The following diagram gives us a better understanding of data type objects:

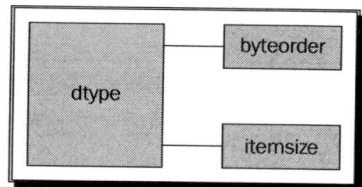

Character codes

Character codes are included for backward compatibility with Numeric. Numeric is the predecessor of NumPy. Their use is not recommended, but the codes are provided here because they pop up in several places. You should instead use `dtype` objects.

Type	Character code
integer	i
Unsigned integer	u
Single precision float	f
Double precision float	d
bool	b
complex	D
string	S
unicode	U
Void	V

Look at the following code to create an array of single precision floats:

```
In: arange(7, dtype='f')
Out: array([ 0.,  1.,  2.,  3.,  4.,  5.,  6.], dtype=float32)
Likewise this creates an array of complex numbers
In: arange(7, dtype='D')
Out: array([ 0.+0.j,  1.+0.j,  2.+0.j,  3.+0.j,  4.+0.j,  5.+0.j,
6.+0.j])
```

dtype constructors

We have a variety of ways to create data types. Take the case of floating point data:

- We can use the general Python float:

```
In: dtype(float)
Out: dtype('float64')
```

- We can specify a single precision float with a character code:

```
In: dtype('f')
Out: dtype('float32')
```

- We can use a double precision float character code:

```
In: dtype('d')
Out: dtype('float64')
```

- We can give the data type constructor a two-character code. The first character signifies the type; the second character is a number specifying the number of bytes in the type:

```
In: dtype('f8')
Out: dtype('float64')
```

A listing of all full data type names can be found in `sctypeDict.keys()`:

```
In: dtype('Float64')
Out: dtype('float64')
```

dtype attributes

The `dtype` class has a number of useful attributes. For example, we can get information about the character code of a data type through the attributes of `dtype`:

```
In: t = dtype('Float64')
In: t.char
Out: 'd'
```

The type attribute corresponds to the type of object of the array elements:

```
In: t.type
Out: <type 'numpy.float64'>
```

The `str` attribute of `dtype` gives a string representation of the data type. It starts with a character representing endianness, if appropriate, then a character code, followed by a number corresponding to the number of bytes that each array item requires. Endianness, here, means the way bytes are ordered within a 32 or 64-bit word. In big-endian order, the most significant byte is stored first. In little-endian order, the least significant byte is stored first.

```
In: t.str
Out: '<f8'
```

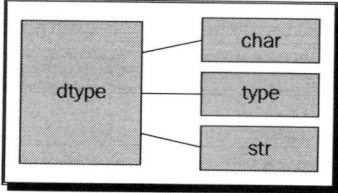

Time for action – creating a record data type

The record data type is a heterogeneous data type—think of it as representing a row in a spreadsheet or a database. To give an example of a record data type, we will create a record for a shop inventory. The record contains the name of the item, a 40-character string, the number of items in the store represented by a 32-bit integer and, finally, a price represented by a 32-bit float. The following steps show how to create a record data type:

1. **Create the record**:

```
In: t = dtype([('name', str_, 40), ('numitems', int32), ('price',
float32)])
In: t
Out: dtype([('name', '|S40'), ('numitems', '<i4'), ('price',
'<f4')])
```

2. **View the type** (we can view the type of a field as well):

```
In: t['name']
Out: dtype('|S40')
```

If you don't give the `array` function a data type, it will assume that it is dealing with floating point numbers. To create the array now, we really have to specify the data type; otherwise, we will get a `TypeError`:

```
In: itemz = array([('Meaning of life DVD', 42, 3.14), ('Butter', 13,
2.72)], dtype=t)
In: itemz[1]
Out: ('Butter', 13, 2.7200000286102295)
```

What just happened?

We created a record data type, which is a heterogeneous data type. The record contained a name as a character string, a number as an integer and a price represented by a float.

One-dimensional slicing and indexing

Slicing of one-dimensional NumPy arrays works just like slicing of Python lists. We can select a piece of an array from index 3 to 7 that extracts the elements 3 through 6:

```
In: a = arange(9)
In: a[3:7]
Out: array([3, 4, 5, 6])
```

We can select elements from index 0 to 7 with a step of 2:

```
In: a[:7:2]
Out: array([0, 2, 4, 6])
```

Similarly as in Python, we can use negative indices and reverse the array:

```
In: a[::-1]
Out: array([8, 7, 6, 5, 4, 3, 2, 1, 0])
```

Time for action – slicing and indexing multidimensional arrays

An ndarray supports slicing over multiple dimensions. For convenience, we refer to many dimensions at once, with an ellipsis.

1. **Create an array and reshape it**: To illustrate, we will create an array with the arange function and reshape it:

```
In: b = arange(24).reshape(2,3,4)
In: b.shape
Out: (2, 3, 4)
In: b
Out:
array([[[ 0,  1,  2,  3],
        [ 4,  5,  6,  7],
        [ 8,  9, 10, 11]],
       [[12, 13, 14, 15],
        [16, 17, 18, 19],
        [20, 21, 22, 23]]])
```

The array b has 24 elements with values 0 to 23 and we reshaped it to be a 2-by-3-by-4, three-dimensional array. We can visualize this as a two-story building with 12 rooms on each floor, 3 rows and 4 columns. As you have probably guessed, the reshape function changes the shape of an array. You give it a tuple of integers, corresponding to the new shape. If the dimensions are not compatible with the data, an exception is thrown.

2. **Selecting a single cell**: We can select a single room by using its three coordinates, namely, the floor, column, and row. For example, the room on the first floor, in the first row, and in the first column (you can have floor 0 and room 0—it's just a matter of convention) can be represented by:

```
In: b[0,0,0]
Out: 0
```

3. **Selecting slices**: If we don't care about the floor, but still want the first column and row, we replace the first index by a : (colon) because we just need to specify the floor number and omit the other indices:

```
In: b[:,0,0]
Out: array([ 0, 12])
This selects the first floor
In: b[0]
Out:
array([[ 0,  1,  2,  3],
       [ 4,  5,  6,  7],
       [ 8,  9, 10, 11]])
```

We could also have written:

```
In: b[0, :, :]
Out:
array([[ 0,  1,  2,  3],
       [ 4,  5,  6,  7],
       [ 8,  9, 10, 11]])
```

An ellipsis replaces multiple colons, so, the preceding code is equivalent to:

```
In: b[0, ...]
Out:
array([[ 0,  1,  2,  3],
       [ 4,  5,  6,  7],
       [ 8,  9, 10, 11]])
```

Further, we get the second row on the first floor with:

```
In: b[0,1]
Out: array([4, 5, 6, 7])
```

4. **Using steps to slice**: Furthermore, we can also select each second element of this selection:

```
In: b[0,1,::2]
Out: array([4, 6])
```

5. **Using ellipsis to slice**: If we want to select all the rooms on both floors that are in the second column, regardless of the row, we will type the following code snippet:

```
In: b[...,1]
Out:
array([[ 1,  5,  9],
       [13, 17, 21]])
```

Similarly, we can select all the rooms on the second row, regardless of floor and column, by writing the following code snippet:

```
In: b[:,1]
Out:
array([[ 4,  5,  6,  7],
       [16, 17, 18, 19]])
```

If we want to select rooms on the ground floor second column, then type the following code snippet:

```
In: b[0,:,1]
Out: array([1, 5, 9])
```

6. **Using negative indices**: If we want to select the first floor, last column, then type the following code snippet:

```
In: b[0,:,-1]
Out: array([ 3,   7,  11])
```

If we want to select rooms on the ground floor, last column reversed, then type the following code snippet:

```
In: b[0,::-1, -1]
Out: array([11,   7,   3])
```

Every second element of that slice:

```
In: b[0,::2,-1]
Out: array([ 3,  11])
```

The command that reverses a one-dimensional array puts the top floor following the ground floor:

```
In: b[::-1]
Out:
array([[[12, 13, 14, 15],
        [16, 17, 18, 19],
        [20, 21, 22, 23]],

       [[ 0,  1,  2,  3],
        [ 4,  5,  6,  7],
        [ 8,  9, 10, 11]]])
```

What just happened?

We sliced a multidimensional NumPy array using several different methods.

Time for action – manipulating array shapes

We already learned about the `reshape` function. Another recurring task is flattening of arrays.

1. **Ravel**: We can accomplish this with the `ravel` function:

```
In: b
Out:
array([[[ 0,  1,  2,  3],
        [ 4,  5,  6,  7],
        [ 8,  9, 10, 11]],

       [[12, 13, 14, 15],
        [16, 17, 18, 19],
        [20, 21, 22, 23]]])
```

```
In: b.ravel()
Out:
array([ 0,  1,  2,  3,  4,  5,  6,  7,  8,  9, 10, 11, 12, 13, 14,
15, 16,
         17, 18, 19, 20, 21, 22, 23])
```

2. **Flatten**: The appropriately-named function, `flatten`, does the same as ravel, but `flatten` always allocates new memory whereas `ravel` might return a view of the array.

```
In: b.flatten()
Out:
array([ 0,  1,  2,  3,  4,  5,  6,  7,  8,  9, 10, 11, 12, 13, 14,
15, 16,
         17, 18, 19, 20, 21, 22, 23])
```

3. **Setting the shape with a tuple**: Besides the `reshape` function, we can also set the shape directly with a tuple, which is shown as follows:

```
In: b.shape = (6,4)
In: b
Out:
array([[ 0,  1,  2,  3],
       [ 4,  5,  6,  7],
       [ 8,  9, 10, 11],
       [12, 13, 14, 15],
       [16, 17, 18, 19],
       [20, 21, 22, 23]])
```

As you can see, this changes the array directly. Now, we have a 6-by-4 array.

4. **Transpose**: In linear algebra, it is common to transpose matrices. We can do that too, by using the following code:

```
In: b.transpose()
Out:
array([[ 0,  4,  8, 12, 16, 20],
       [ 1,  5,  9, 13, 17, 21],
       [ 2,  6, 10, 14, 18, 22],
       [ 3,  7, 11, 15, 19, 23]])
```

5. **Resize**: The `resize` melthod works just like the `reshape` method, but modifies the array it operates on:

```
In: b.resize((2,12))
In: b
Out:
array([[ 0,  1,  2,  3,  4,  5,  6,  7,  8,  9, 10, 11],
       [12, 13, 14, 15, 16, 17, 18, 19, 20, 21, 22, 23]])
```

What just happened?

We manipulated the shapes of NumPy arrays using the `ravel` function, function `flatten`, the `reshape` function, and the `resize` method.

Stacking

Arrays can be stacked horizontally, depth-wise, or vertically. We can use, for that purpose, the `vstack`, `dstack`, `hstack`, `column_stack`, `row_stack`, and `concatenate` functions.

Time for action – stacking arrays

First, let's set up some arrays:

```
In: a = arange(9).reshape(3,3)
In: a
Out:
array([[0, 1, 2],
       [3, 4, 5],
       [6, 7, 8]])
In: b = 2 * a
In: b
Out:
array([[ 0,  2,  4],
       [ 6,  8, 10],
       [12, 14, 16]])
```

1. **Horizontal stacking**: Starting with horizontal stacking, we will form a tuple of ndarrays and give it to the `hstack` function. This is shown as follows:

    ```
    In: hstack((a, b))
    Out:
    array([[0, 1, 2, 0, 2, 4],
           [3, 4, 5, 6, 8, 10],
           [6, 7, 8, 12, 14, 16]])
    ```

    We can achieve the same with the `concatenate` function, which is shown as follows:

    ```
    In: concatenate((a, b), axis=1)
    Out:
    array([[ 0, 1, 2, 0, 2, 4],
           [ 3, 4, 5, 6, 8, 10],
           [ 6, 7, 8, 12, 14, 16]])
    ```

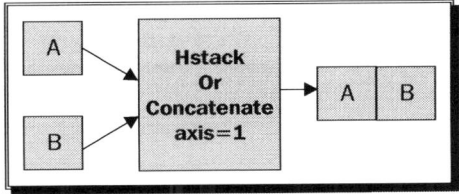

2. **Vertical stacking**: With vertical stacking, again, a tuple is formed. This time, it is given to the vstack function. This can be seen as follows:

```
In: vstack((a, b))
Out:
array([[ 0,   1,   2],
       [ 3,   4,   5],
       [ 6,   7,   8],
       [ 0,   2,   4],
       [ 6,   8,  10],
       [12,  14,  16]])
```

The concatenate function produces the same result with the axis set to 0. This is the default value for the axis argument.

```
In: concatenate((a, b), axis=0)
Out:
array([[ 0,   1,   2],
       [ 3,   4,   5],
       [ 6,   7,   8],
       [ 0,   2,   4],
       [ 6,   8,  10],
       [12,  14,  16]])
```

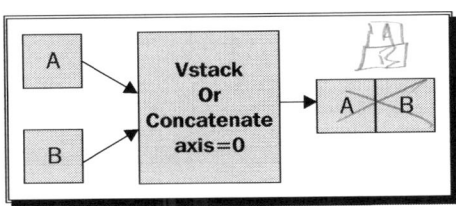

3. **Depth stacking**: Additionally, there is the depth-wise stacking using dstack and a tuple, of course. This means stacking of a list of arrays along the third axis (depth). For instance, we could stack 2D arrays of image data on top of each other.

```
In: dstack((a, b))
Out:
array([[[ 0,   0],
```

```
      [ 1,   2],
      [ 2,   4]],
     [[ 3,   6],
      [ 4,   8],
      [ 5, 10]],
     [[ 6, 12],
      [ 7, 14],
      [ 8, 16]]])
```

4. **Column stacking**: The `column_stack` function stacks 1D arrays column-wise. It's shown as follows:

```
In: oned = arange(2)
In: oned
Out: array([0, 1])
In: twiceoned = 2 * oned
In: twiceoned
Out: array([0, 2])
In: column_stack((oned, twiceoned))
Out:
array([[0, 0],
       [1, 2]])
```

2D arrays are stacked the way `hstack` stacks them:

```
In: column_stack((a, b))
Out:
array([[ 0,  1,  2,  0,  2,  4],
       [ 3,  4,  5,  6,  8, 10],
       [ 6,  7,  8, 12, 14, 16]])
In: column_stack((a, b)) == hstack((a, b))
Out:
array([[ True,  True,  True,  True,  True,  True],
       [ True,  True,  True,  True,  True,  True],
       [ True,  True,  True,  True,  True,  True]], dtype=bool)
```

Yes, you guessed it right! We compared two arrays with the `==` operator. Isn't it beautiful?

5. **Row stacking**: NumPy, of course, also has a function that does row-wise stacking. It is called `row_stack` and, for 1D arrays, it just stacks the arrays in rows into a 2D array.

```
In: row_stack((oned, twiceoned))
Out:
array([[0, 1],
       [0, 2]])
```

The `row_stack` function results for 2D arrays are equal to. Yes, exactly the `vstack` function results.

```
In: row_stack((a, b))
Out:
array([[ 0,   1,   2],
       [ 3,   4,   5],
       [ 6,   7,   8],
       [ 0,   2,   4],
       [ 6,   8,  10],
       [12,  14,  16]])
In: row_stack((a,b)) == vstack((a, b))
Out:
array([[ True,   True,   True],
       [ True,   True,   True],
       [ True,   True,   True],
       [ True,   True,   True],
       [ True,   True,   True],
       [ True,   True,   True]], dtype=bool)
```

What just happened?

We stacked arrays horizontally, depth-wise, or vertically. We used the `vstack`, `dstack`, `hstack`, `column_stack`, `row_stack`, and `concatenate` functions.

Splitting

Arrays can be split vertically, horizontally, or depth wise. The functions involved are `hsplit`, `vsplit`, `dsplit`, and `split`. We can either split into arrays of the same shape or indicate the position after which the split should occur.

Time for action – splitting arrays

1. **Horizontal splitting**: The ensuing code splits an array along its horizontal axis into three pieces of the same size and shape. This is shown as follows:

```
In: a
Out:
array([[0, 1, 2],
       [3, 4, 5],
       [6, 7, 8]])
In: hsplit(a, 3)
Out:
[array([[0],
        [3],
```

```
       [6]]),
array([[1],
       [4],
       [7]]),
array([[2],
       [5],
       [8]])]
```

Compare it with a call of the `split` function, with extra parameter `axis=1`:

```
In: split(a, 3, axis=1)
Out:
[array([[0],
       [3],
       [6]]),
array([[1],
       [4],
       [7]]),
array([[2],
       [5],
       [8]])]
```

2. **Vertical splitting**: `vsplit` splits along the vertical axis: *— ie. splits to individual rows.*

```
In: vsplit(a, 3)
Out: [array([[0, 1, 2]]), array([[3, 4, 5]]), array([[6, 7, 8]])]
```

The `split` function, with `axis=0`, also splits along the vertical axis:

```
In: split(a, 3, axis=0)
Out: [array([[0, 1, 2]]), array([[3, 4, 5]]), array([[6, 7, 8]])]
```

3. **Depth-wise splitting**: The `dsplit` function, unsurprisingly, splits depth-wise. We will need an array of rank 3 first:

```
In: c = arange(27).reshape(3, 3, 3)
In: c
Out:
array([[[ 0,  1,  2],
        [ 3,  4,  5],
        [ 6,  7,  8]],
       [[ 9, 10, 11],
        [12, 13, 14],
        [15, 16, 17]],
       [[18, 19, 20],
        [21, 22, 23],
        [24, 25, 26]]])
```

```
In: dsplit(c, 3)
Out:
[array([[[  0],
         [  3],
         [  6]],
        [[  9],
         [ 12],
         [ 15]],
        [[ 18],
         [ 21],
         [ 24]]]),
 array([[[  1],
         [  4],
         [  7]],
        [[ 10],
         [ 13],
         [ 16]],
        [[ 19],
         [ 22],
         [ 25]]]),
 array([[[  2],
         [  5],
         [  8]],
        [[ 11],
         [ 14],
         [ 17]],
        [[ 20],
         [ 23],
         [ 26]]])]
```

What just happened?

We split arrays using the `hsplit`, `vsplit`, `dsplit`, and `split` functions.

Array attributes

Besides the `shape` and `dtype` attributes, `ndarray` has a number of other attributes, as shown in the following list:

◆ `ndim` gives the number of dimensions:

```
In: b
Out:
array([[  0,   1,   2,   3,   4,   5,   6,   7,   8,   9,  10,  11],
```

```
        [12, 13, 14, 15, 16, 17, 18, 19, 20, 21, 22, 23]])
In: b.ndim
Out: 2
```

◆ `size` contains the number of elements. This is shown a follows:

```
In: b.size
Out: 24
```

◆ `itemsize` gives the number of bytes for each element in the array:

```
In: b.itemsize

Out: 8
```

◆ If you want the total number of bytes the array requires, you can have a look at nbytes. This is just a product of the `itemsize` and `size` attributes:

```
In: b.nbytes
Out: 192
In: b.size * b.itemsize
Out: 192
```

◆ The `T` attribute has the same effect as the `transpose` function, which is shown as follows:

```
In: b.resize(6,4)
In: b
Out:
array([[ 0,  1,  2,  3],
       [ 4,  5,  6,  7],
       [ 8,  9, 10, 11],
       [12, 13, 14, 15],
       [16, 17, 18, 19],
       [20, 21, 22, 23]])
In: b.T
Out:
array([[ 0,  4,  8, 12, 16, 20],
       [ 1,  5,  9, 13, 17, 21],
       [ 2,  6, 10, 14, 18, 22],
       [ 3,  7, 11, 15, 19, 23]])
```

◆ If the array has a rank lower than 2, we will just get a view of the array:

```
In: b.ndim
Out: 1
In: b.T
Out: array([0, 1, 2, 3, 4])
```

Complex numbers in NumPy are represented by .j. For example, we can create an array with complex numbers:

```
In: b = array([1.j + 1, 2.j + 3])
In: b
Out: array([ 1.+1.j,   3.+2.j])
```

◆ The `real` attribute gives us the real part of the array, or the array itself if it only contains real numbers:

```
In: b.real
Out: array([ 1.,   3.])
```

◆ The `imag` attribute contains the imaginary part of the array:

```
In: b.imag
Out: array([ 1.,   2.])
```

◆ If the array contains complex numbers, then the data type is automatically also complex:

```
In: b.dtype
Out: dtype('complex128')
In: b.dtype.str
Out: '<c16'
```

◆ The `flat` attribute returns a `numpy.flatiter` object. This is the only way to acquire a `flatiter`—we do not have access to a `flatiter` constructor. The flat iterator enables us to loop through an array as if it is a flat array, as shown next:

```
In: b = arange(4).reshape(2,2)
In: b
Out:
array([[0, 1],
       [2, 3]])
In: f = b.flat
In: f
Out: <numpy.flatiter object at 0x103013e00>
In: for item in f: print item
    .....:
0
1
2
3
```

It is possible to directly get an element with the `flatiter` object:

```
In: b.flat[2]
Out: 2
```

or multiple elements:

```
In: b.flat[[1,3]]
Out: array([1, 3])
```

The `flat` attribute is settable. Setting the value of the `flat` attribute leads to overwriting the values of the whole array:

```
In: b.flat = 7
In: b
Out:
array([[7, 7],
       [7, 7]])
or selected elements
In: b.flat[[1,3]] = 1
In: b
Out:
array([[7, 1],
       [7, 1]])
```

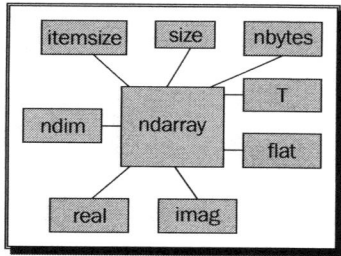

Time for action – converting arrays

We can convert a NumPy array to a Python list with the `tolist` function. This is shown as follows:

1. **Convert to a list**:

   ```
   In: b
   Out: array([ 1.+1.j,  3.+2.j])
   In: b.tolist()
   Out: [(1+1j), (3+2j)]
   ```

2. **astype function**: The `astype` function converts the array to an array of the specified type:

   ```
   In: b
   Out: array([ 1.+1.j,  3.+2.j])
   ```

```
In: b.astype(int)
/usr/local/bin/ipython:1: ComplexWarning: Casting complex values
to real discards the imaginary part
  #!/usr/bin/python
Out: array([1, 3])
```

 We are losing the imaginary part when casting from complex type to int. The astype function also accepts the name of a type as a string.

```
In: b.astype('complex')
Out: array([ 1.+1.j,   3.+2.j])
```

It won't show any warning this time, because we used the proper data type.

What just happened?

We converted NumPy arrays to a list and to arrays of different data types.

Summary

We learned a lot in this chapter about the NumPy fundamentals: data types and arrays. Arrays have several attributes describing them. We learned that one of these attributes is the data type, which, in NumPy, is represented by a full-fledged object.

NumPy arrays can be sliced and indexed in an efficient manner, just like Python lists. NumPy arrays have the added ability of working with multiple dimensions.

The shape of an array can be manipulated in many ways—stacking, resizing, reshaping, and splitting. A great number of convenience functions for shape manipulation were demonstrated in this chapter.

Having learned about the basics, it's time to move on to the study of commonly-used functions in *Chapter 3, Get to terms with commonly used functions*. This includes basic statistical and mathematical functions.

3
Get into Terms with Commonly Used Functions

In this chapter, we will have a look at common NumPy functions. In particular, we will learn how to load data from files using a historical stock prices example. Also, we will get to see the basic NumPy mathematical and statistical functions.

We will learn how to read from, and write to, files. Also, we will get a taste of the functional programming and linear algebra possibilities in NumPy.

In this chapter, we shall cover the following topics:

- ◆ Functions working on arrays
- ◆ Loading arrays from files
- ◆ Writing arrays to files
- ◆ Simple mathematical and statistical functions

File I/O

First, we will learn about file I/O with NumPy. Data is usually stored in files. You would not get far if you are not able to read from and write to files.

Time for action – reading and writing files

As an example of file I/O, we will create an identity matrix and store its contents in a file.

Identity matrix creation

1. **Creating an identity matrix**: The identty matrix is a square matrix with ones on the diagonal and zeroes for the rest.

The **identity matrix** can be created with the `eye` function. The only argument we need to give the `eye` function is the number of ones. So, for instance, for a 2-by-2 matrix, write the following code:

```
i2 = numpy.eye(2)
print i2
```

The output is:

```
[[ 1.   0.]
 [ 0.   1.]]
```

2. **Saving data**: Save the data with the `savetxt` function. We obviously need to specify the name of the file that we want to save the data in and the array containing the data itself:

```
numpy.savetxt("("eye.txt", i2)
```

A file called `eye.txt` should have been created. You can check for yourself whether the contents are as expected.

What just happened?

Reading and writing files is a necessary skill for data analysis. We wrote to a file with `savetxt`. We made an identity matrix with the `eye` function.

CSV files

Files in the **comma separated values (CSV)** format are encountered quite frequently. Often, the CSV file is just a dump from a database file. Usually, each field in the CSV file corresponds to a database table column. As we all know, spreadsheet programs, such as Excel, can produce CSV files as well.

Time for action – loading from CSV files

How do we deal with CSV files? Luckily, the `loadtxt` function can conveniently read CSV files, split up the fields and load the data into NumPy arrays. In the following example, we will load historical price data for Apple (the company, not the fruit). The data is in CSV format. The first column contains a symbol that identifies the stock. In our case, it is AAPL, next in our case. Nn is the date in dd-mm-yyyy format. The third column is empty. Then, in order, we have the open, high, low, and close price. Last, but not least, is the volume of the day. This is what a line looks like:

```
AAPL,28-01-2011, ,344.17,344.4,333.53,336.1,21144800
```

- ◆ **Loading data**: For now, we are only interested in the close price and volume. In the preceding sample, that would be 336.1 and 21144800. Store the close price and volume in two arrays as follows:

```
c,v=numpy.loadtxt('data.csv', delimiter=',', usecols=(6,7),
unpack=True)))
```

As you can see, data is stored in the `data.csv` file. We have set the delimiter to **, (comma)**, since we are dealing with a comma separated value file. The `usecols` parameter is set through a tuple to get the seventh and eighth fields, which correspond to the close price and volume. `Unpack` is set to `True`, which means that data will be unpacked and assigned to the `c` and `v` variables that will hold the close price and volume, respectively.

What just happened?

CSV files are a special type of file that we have to deal with frequently. We read a CSV file containing stock quotes with the `loadtxt` function. We indicated to the `loadtxt` function that the delimiter of our file was a comma. We specified which columns we were interested in, through the `usecols` argument, and set the `unpack` parameter to `True` so that the data was unpacked for further use.

Volume weighted average price

Volume weighted average price (VWAP) is a very important quantity. The higher the volume, the more significant a price move typically is. VWAP is calculated by using volume values as weights.

Time for action – calculating volume weighted average price

These are the actions that we will take:

1. Read the data into arrays.

2. Calculate VWAP:

```
import numpy
c,v=numpy.loadtxt('data.csv', delimiter=',', usecols=(6,7),
unpack=True)
vwap = numpy.average(c, weights=v)
print "VWAP =", vwap
The output is
VWAP = 350.589549353
```

What just happened?

That wasn't very hard, was it? We just called the `average` function and set its `weights` parameter to use the `v` array for weights. By the way, NumPy also has a function to calculate the arithmetic mean.

The mean function

The `mean` function is quite friendly and not so mean. This function calculates the arithmetic mean of an array. Let's see it in action:

```
print "mean =", numpy.mean(c)
mean =   351.037666667
```

Time weighted average price

Now that we are at it, let's compute the time weighted average price too. It is just a variation on a theme really. The idea is that recent price quotes are more important, so we should give recent prices higher weights. The easiest way is to create an array with the `arange` function of increasing values from zero to the number of elements in the close price array. This is not necessarily the correct way. In fact, most of the examples concerning stock price analysis in this book are only illustrative. The following is the TWAP code:

```
t = numpy.arange(len(c))
print "twap =", numpy.average(c, weights=t)
```

It produces this output:

```
twap = 352.428321839
```

The TWAP is even higher than the mean.

Pop quiz – computing the weighted average

1. Which function returns the weighted average of an array?

   a. weighted average

   b. waverage

   c. average — *with weights*

   d. avg

Have a go hero – calculating other averages

Try doing the same calculation using the open price. Calculate the mean for the volume and the other prices.

Value range

Usually, we don't only want to know the average or arithmetic mean of a set of values, which are sort of in the middle; we also want the extremes, the full range—the highest and lowest values. The sample data that we are using here already has those values per day—the high and low price. However, we need to know the highest value of the high price and the lowest price value of the low price. After all, how else would we know how much our Apple stocks would gain or lose.

Time for action – finding highest and lowest values

The `min` and `max` functions are the answer to our requirement.

1. **Reading from a file**: First, we will need to read our file again and store the values for the high and low prices into arrays:

   ```
   h,l=numpy.loadtxt('data.csv', delimiter=',', usecols=(4,5),
   unpack=True)
   ```

   The only thing that changed is the `usecols` parameter, since the high and low prices are situated in different columns.

2. **Getting the range**: The following code gets the price range:

   ```
   print "highest =", numpy.max(h)))
   print "lowest =", numpy.min(l)
   ```

   These are the values returned:

   ```
   highest = 364.9
   lowest = 333.53
   ```

   Now, it's trivial to get a midpoint, so it is left as an exercise for the reader to attempt.

3. **Calculating the spread**: NumPy allows us to compute the spread of an array with a function called The `ptp` function returns the difference between the maximum and minimum values of an array. In other words, it is equal to `max(array) - min(array)`. Call the `ptp` function:

```
print "Spread high price", numpy.ptp(h)
print "Spread low price", numpy.ptp(l)
```
peak to peak.

You will see this:

```
Spread high price 24.86
Spread low price 26.97
```

What just happened?

We defined a range of highest to lowest values for the price. The highest value was given by applying the `max` function to the high price array. Similarly, the lowest value was found by calling the `min` function to the low price array. We also calculated the peak to peak distance with the ptp function.

Statistics

Stock traders are interested in the most probable close price. Common sense says that this should be close to some kind of an average. The arithmetic mean and weighted average are ways to find the center of a distribution of values. However, both are not robust and sensitive to outliers. For instance, if we had a close price value of a million dollars, this would have influenced the outcome of our calculations.

Time for action – doing simple statistics

One thing that we can do is use some kind of threshold to weed out outliers, but there is a better way. It is called the median, and it basically picks the middle value of a sorted set of values. For example, if we have the values of 1, 2, 3, 4 and 5. The median would be 3, since it is in the middle. These are the steps to calculate the median:

1. **Determine the median of the close price**: Create a new Python script and call it `simplestats.py`. You already know how to load the data from a CSV file into an array. So, copy that line of code and make sure that it only gets the close price. The code should appear like this, by now:

```
c=numpy.loadtxt('data.csv', delimiter=',', usecols=(6,),
unpack=True)
```

The function that will do the magic for us is called median. We will call it and print the result immediately. Add the following line of code:

```
print "median =", numpy.median(c)
```

The program prints the following output:

```
median = 352.055
```

Since it is our first time using the median function, we would like to check whether this is correct. Not because we are paranoid or anything! Obviously, we could do it by just going through the file and finding the correct value, but that is no fun. Instead we will just mimic the median algorithm by sorting the close price array and printing the middle value of the sorted array. The msort function does the first part for us. We will call the function, store the sorted array, and then print it:

```
sorted_close = numpy.msort(c)
print "sorted =", sorted_close
```

This prints the following output:

```
sorted = [ 336.1    338.61   339.32   342.62   342.88   343.44   344.32
345.03   346.5
  346.67   348.16   349.31   350.56   351.88   351.99   352.12   352.47
353.21
  354.54   355.2    355.36   355.76   356.85   358.16   358.3    359.18
359.56
  359.9    360.     363.13]
```

Yup, it works! Let's now get the middle value of the sorted array:

```
N = len(c)
print "middle =", sorted[(N - 1)/2]
```

It gives us the following output:

```
middle = 351.99
```

Hey, that's a different value than the one the median function gave us. How come? Upon further investigation we find that the median function return value doesn't even appear in our file. That's even stranger! Before filing bugs with the NumPy team, let's have a look at the documentation. This mystery is easy to solve. It turns out that our naive algorithm only works for arrays with odd lengths. For even-length arrays, the median is calculated from the average of the two array values in the middle. Therefore, type the following code:

```
print "average middle =", (sorted[N /2] + sorted[(N - 1) / 2]) / 2
```

This prints the following output:

```
average middle = 352.055
```

Success!

Another statistical measure that we are concerned with is **variance**. Variance tells us how much a variable varies. In our case, it also tells us how risky an investment is, since a stock price that varies too wildly is bound to get us into trouble.

2. **Calculate the variance of the close price**: With NumPy, this is just a one liner. See the following code:

```
print "variance =", numpy.var(c)
```

This gives us the following output:

```
variance = 50.1265178889
```

Not that we don't trust NumPy or anything, but let's double-check using the definition of variance, as found in the documentation. Mind you, this definition might be different than the one in your statistics book, but that is quite common in the field of statistics. The variance is defined as the mean of the square of deviations from the mean, divided by the number of elements in the array. Some books tell us to divide by the number of elements in the array minus one.

```
print "variance from definition =", numpy.mean((c - c.mean())**2)
```

The output is as follows:

```
variance from definition = 50.1265178889
```

Just as we expected!

What just happened?

Maybe you noticed something new. We suddenly called the mean function on the c array. Yes, this is legal, because the ndarray object has a mean method. This is for your convenience. For now, just keep in mind that this is possible.

Stock returns

In academic literature it is more common to base analysis on stock returns and log returns of the close price. Simple returns are just the rate of change from one value to the next. Logarithmic returns or log returns are determined by taking the log of all the prices and calculating the differences between them. In high school, we learned that the difference between the log of "a" and the log of "b" is equal to the log of "a divided by b". Log return, therefore, also measures rate of change. Returns are dimensionless, since, in the act of dividing, we divide dollar by dollar (or some other currency). Anyway, investors are most likely to be interested in the variance or standard deviation of the returns, as this represents risk.

Time for action – analyzing stock returns

Follow the ensuing steps to analyze stock returns:

1. **Simple returns**: First, let's calculate simple returns. NumPy has the `diff` function that returns an array built up of the difference between two consecutive array elements. This is sort of like differentiation in calculus. To get the returns, we also have to divide by the value of the previous day. We must be careful though. The array returned by `diff` is one element shorter than the close prices array. After careful deliberation, we get the following code:

   ```
   returns = numpy.diff( arr ) / arr[ : -1]
   ```
   dist between consecutive values

   Notice that we don't use the last value in the divisor. Let's compute the standard deviation using the `std` function:

   ```
   print "Standard deviation =", numpy.std(returns)
   ```

   This results in the following output:

   ```
   Standard deviation = 0.0129221344368
   ```

2. **Logarithmic returns**: The log return is even easier to calculate. We use the `log` function to get the log of the close price and then unleash the `diff` function on the result. This is shown as follows:

   ```
   logreturns = numpy.diff( numpy.log(c) )
   ```

   Normally, we would have to check that the input array doesn't have zeroes or negative numbers. If it did we would have gotten an error. Stock prices are, however, always positive, so we didn't have to check.

3. **Selecting positive returns**: Quite likely, we will be interested in days when the return is positive. In the current setup, we can get the next best thing with the `where` function, which returns the indices of an array that satisfies a condition. Just type the following code:

   ```
   posretindices = numpy.where(returns > 0)
   print "Indices with positive returns", posretindices
   ```

   This gives us a number of indices for the array elements that are positive:

   ```
   Indices with positive returns (array([ 0,  1,  4,  5,  6,  7,  9,
   10, 11, 12, 16, 17, 18, 19, 21, 22, 23, 25, 28]),)
   ```

4. **Annualized and monthly volatilities**: In investing, volatility measures price variation of a financial security. Historical volatility is calculated from historical price data. The logarithmic returns are interesting if you want to know the historical volatility—for instance, the annualized or monthly volatility. The annualized volatility is equal to the standard deviation of the log returns as a ratio of its mean, divided by one over the square root of the number of business days in a year, usually one assumes 252. Calculate it with the `std` and `mean` functions. See the following code:

```
annual_volatility = numpy.std(logreturns)/numpy.mean(logreturns)
annual_volatility = annual_volatility / numpy.sqrt(1./252.).)
print annual_volatility
```

Take notice of the division within the `sqrt` function. Since, in Python, integer division works differently than float division, we needed to use floats to make sure that we get the proper results. The monthly volatility is similarly given by:

```
print "Monthly volatility", annual_volatility * numpy.sqrt(1./12.)
```

What just happened?

We calculated the simple stock returns with the `diff` function, which calculates differences between sequential elements. The `log` function computes the natural logarithms of array elements. We used it to calculate the logarithmic returns. At the end of the tutorial we calculated the annual and monthly volatility.

Dates

Do you sometimes have the Monday blues or the Friday fever? Ever wondered whether the stock market suffers from said phenomena? Well, I think this certainly warrants extensive research.

Time for action – dealing with dates

First, we will read the close price data. Second, we will split the prices according to the day of the week. Third, for each weekday, we will calculate the average price. Finally, we will find out which day of the week has the highest average and which has the lowest average. A health warning before we commence: you might be tempted to use the result to buy stock on one day and sell on the other. However, we don't have enough data to make these kind of decisions. Please consult a professional statistician first!

Coders hate dates because they are so complicated! NumPy is very much oriented towards floating point operations. For that reason, we need to take extra effort to process dates. Try it out yourself; put the following code in a script or use the one that comes with the book:

```
dates, close=numpy.loadtxt('data.csv', delimiter=',',
    usecols=(1,6), unpack=True)
```

Execute the script and the following error will appear:

```
ValueError: invalid literal for float(): 28-01-2011
```

1. **Converter function**: Obviously, NumPy tried to convert the dates into floats. What we have to do is explicitly tell NumPy how to convert the dates. The `loadtxt` function has a special parameter for this purpose. The parameter is called `converters` and is a dictionary that links columns with so-called converter functions. It is our responsibility to write the converter function.

   Let's write the function down:

   ```
   # Monday 0
   # Tuesday 1
   # Wednesday 2
   # Thursday 3
   # Friday 4
   # Saturday 5
   # Sunday 6
   def datestr2num(s):
       return datetime.datetime.strptime(s, "%d-%m-%Y").date().
   weekday()
   ```

   We give the `datestr2num` function dates as a string, such as "28-01-2011". The string is first turned into a `datetime` object using a specified format `"%d-%m-%Y"`. This is, by the way, standard Python and is not related to NumPy itself. Second, the `datetime` object is turned into a day. Finally the `weekday` method is called on the date to return a number. As you can read in the comments, the number is between 0 and 6. 0 is for instance Monday and 6 is Sunday. The actual number, of course, is not important for our algorithm; it is only used as identification.

2. **Load the data**: Now we will hook up our date converter function:

   ```
   dates, close=numpy.loadtxt('data.csv', delimiter=',',
   usecols=(1,6), converters={1: datestr2num}, unpack=True)
   print "Dates =", dates
   ```

   This prints the following output:

   ```
   Dates = [ 4.  0.  1.  2.  3.  4.  0.  1.  2.  3.  4.  0.  1.  2.
   3.  4.  1.  2.  4.  0.  1.  2.  3.  4.  0.  1.  2.  3.  4.]
   ```

   No Saturdays and Sundays, as you can see. Exchanges are closed over the weekend.

3. **Initialize the averages array**: We will now make an array that has five elements for each day of the week. The values of the array will be initialized to 0:

```
averages = numpy.zeros(5)
```

This array will hold the averages for each weekday.

4. **Calculate the averages**: We already learned about the `where` function that returns indices of the array for elements that conform to a specified condition. The `take` function can use these indices and takes the values of the corresponding array items. We will use the `take` function to get the close prices for each week day. In the following loop we go through the date values which are 0 to 4, better known as Monday to Friday. We get the indices with the `where` function for each day and store it in the `indices` array. Then, we retrieve the values corresponding to the indices, using the `take` function. Finally we compute an average for each weekday and store it in the `averages` array, like so:

```
for i in range(5):
    indices = numpy.where(dates == i)
    prices = numpy.take(close, indices)
    avg = numpy.mean(prices)
    print "Day", i, "prices", prices, "Average", avg
    averages[i] = avg
```

The loop prints the following output:

```
Day 0 prices [[ 339.32  351.88  359.18  353.21  355.36]] Average
351.79
Day 1 prices [[ 345.03  355.2   359.9   338.61  349.31  355.76]]
Average 350.635
Day 2 prices [[ 344.32  358.16  363.13  342.62  352.12  352.47]]
Average 352.136666667
Day 3 prices [[ 343.44  354.54  358.3   342.88  359.56  346.67]]
Average 350.898333333
Day 4 prices [[ 336.1   346.5   356.85  350.56  348.16  360.
351.99]] Average 350.022857143
```

5. **Find the maxima and minima**: If you want, you can go ahead and find out which day has the highest, and which the lowest, average. However, it is just as easy to find this out with the `max` and `min` functions, as shown here:

```
top = numpy.max(averages)
print "Highest average", top
print "Top day of the week",  numpy.argmax(averages)
bottom = numpy.min(averages)
print "Lowest average", bottom
print "Bottom day of the week",  numpy.argmin(averages)
```

The output is as follows:

```
Highest average 352.136666667
Top day of the week 2
Lowest average 350.022857143
Bottom day of the week 4
```

What just happened?

The `argmin` function returned the index of the lowest value in the `averages` array. The index returned was 4, which corresponds to Friday. The `argmax` function returned the index of the highest value in the `averages` array. The index returned was 2, which corresponds to Wednesday.

Have a go hero – looking at VWAP and TWAP

Hey, that was fun! For the sample data, it appears that Friday is the cheapest day and Wednesday is the day when your Apple stock will be worth the most. Ignoring the fact that we have very little data, is there a better method to compute the averages? Shouldn't we involve volume data as well? Maybe it makes more sense to you to do a time-weighted average. Give it a go! Calculate the VWAP and TWAP. You can find some hints on how to go about doing this at the beginning of this chapter.

Weekly summary

The data that we used in the previous *Time for action* tutorials is end-of-day data. In essence, it is summarized data compiled from trade data for a certain day. If you are interested in the cotton market and have decades of data, you might want to summarize and compress the data even further. Let's do that. Let's summarize the data of Apple stocks to give us weekly summaries.

Time for action – summarizing data

The data we will summarize will be for a whole business week from Monday to Friday. During the period covered by the data, there was one holiday on February 21st, President's Day. This happened to be a Monday and the US stock exchanges were closed on this day. As a consequence, there is no entry for this day, in the sample. The first day in the sample is a Friday, which is inconvenient. Use the following instructions to summerize data:

1. **Selecting the first three weeks**: To simplify, we will just have a look at the first three weeks in the sample—you can later have a go at improving this:

   ```
   close = close[:16]
   dates = dates[:16]
   ```

   We will build on the code from the previous *Time for action* tutorial.

2. **Finding the first Monday**: Commencing, we will find the first Monday in our sample data. Recall that Mondays have the code 0 in Python. This is what we will put in the condition of a `where` function. Then, we will need to extract the first element that has index 0. The result would be a multidimensional array. Flatten that with the `ravel` function:

```
# get first Monday
first_monday = numpy.ravel(numpy.where(dates == 0))[0]
print "The first Monday index is", first_monday
```

This will print the following output:

```
The first Monday index is 1
```

3. **Finding the last Friday**: The next logical step is to find the Friday before last Friday in the sample. The logic is similar to the one for finding the first Monday, and the code for Friday is 4. Additionally, we are looking for the second-to-last element with index 2.

```
# get last Friday
last_friday = numpy.ravel(numpy.where(dates == 4))[-2]
print "The last Friday index is", last_friday
```

This will give us the following output:

```
The last Friday index is 15
```

Creating arrays with multi-week indices: Next, create an array with the indices of all the days in the three weeks

```
weeks_indices = numpy.arange(first_monday, last_friday + 1)
print "Weeks indices initial", weeks_indices
```

4. **Splitting the array**: Split the array in pieces of size 5 with the `split` function.

```
weeks_indices = numpy.split(weeks_indices, 5)
print "Weeks indices after split", weeks_indices
```

It splits the array as follows:

```
Weeks indices after split [array([1, 2, 3, 4, 5]), array([ 6,  7,
8,  9, 10]), array([11, 12, 13, 14, 15])]
```

5. **Calling the apply_along_axis function**: In NumPy, dimensions are called axes. Now, we will get fancy with the `apply_along_axis` function. This function calls another function, which we will provide, to operate on each of the elements of an array. Currently, we have an array with three elements. Each array item corresponds to one week in our sample and contains indices of the corresponding items. Call the `apply_along_axis` function by supplying the name of our function, called `summarize,` that we will define shortly. Further specify the axis or dimension number (such as 1), the array to operate on, and a variable number of arguments for the `summarize` function, if any:

```
weeksummary = numpy.apply_along_axis(summarize, 1, weeks_indices,
open, high, low, close)
print "Week summary", weeksummary
```

6. **Write the summarize function**: The `summarize` function returns, for each week, a tuple that holds the open, high, low, and close price for the week, similarly to end-of-day data:

```
def summarize(a, o, h, l, c):
    monday_open = o[a[0]]
    week_high = numpy.max( numpy.take(h, a) )
    week_low = numpy.min( numpy.take(l, a) )
    friday_close = c[a[-1]]

    return("APPL", monday_open, week_high, week_low, friday_close)
```

Notice that we used the `take` function to get the actual values from indices. Calculating the high and low values of the week was easily done with the `max` and `min` functions. The `open` for the week is the open for the first day in the week—Monday. Likewise, the `close` is the close for the last day of the week—Friday:

```
Week summary [['APPL' '335.8' '346.7' '334.3' '346.5']
 ['APPL' '347.89' '360.0' '347.64' '356.85']
 ['APPL' '356.79' '364.9' '349.52' '350.56']]
```

7. **Writing the date to a file**: Store the data in a file with the NumPy `savetxt` function:

```
numpy.savetxt("weeksummary.csv", weeksummary, delimiter=",",
fmt="%s")
```

As you can see, we specify a filename, the array we want to store, a delimiter (in this case a comma), and the format we want to store floating point numbers in.

The format string starts with a percent sign. Second is an optional flag. The—flag means left justify, 0 means left pad with zeroes, + means precede with + or -. Third is an optional width. The width indicates the minimum number of characters. Fourth, a dot is followed by a number linked to precision. Finally, there comes a character specifier; in our example, the character specifier is a string.

Character code	Description
c	character
d or i	signed decimal integer
e or E	scientific notation with e or E.
f	decimal floating point
g,G	use the shorter of e,E or f
o	signed octal
s	string of characters
u	unsigned decimal integer
x,X	unsigned hexadecimal integer

View the generated file in your favorite editor or type at the command line:

```
cat weeksummary.csv
APPL,335.8,346.7,334.3,346.5
APPL,347.89,360.0,347.64,356.85
APPL,356.79,364.9,349.52,350.56
```

What just happened?

We did something that is not even possible in some programming languages. We defined a function and passed it as an argument to the `apply_along_axis` function. Arguments for the `summarize` function were neatly passed by `apply_along_axis`.

Have a go hero – improving the code

Change the code to deal with a holiday. Time the code to see how big the speedup due to `apply_along_axis` is.

Average true range

The **average true range (ATR)** is a technical indicator that measures volatility of stock prices. The ATR calculation is not important further but will serve as an example of several NumPy functions, including the `maximum` function.

Time for action – calculating the average true range

To calculate the average true range, follow the ensuing steps:

1. **Selecting the last N days**: The ATR is based on the low and high price of N days, usually the last 20 days.

```
N = int(sys.argv[1])
h = h[-N:]
l = l[-N:]
```

2. **Retrieving the previous close days price**: We also need to know the close price of the previous day:

```
previousclose = c[-N -1: -1]
```

For each day, we calculate the following:

The daily range—the difference between high and low price:

```
h - l
```

The difference between high and previous close:

```
h - previousclose
```

The difference between the previous close and the low price:

```
previousclose - l
```

3. **Computing the true range**: The `max` function returns the maximum of an array. Based on those three values, we calculate the so-called true range, which is the maximum of these values. We are now interested in the element-wise maxima across arrays – meaning the maxima of the first elements in the arrays, the second elements in the arrays, and so on. Use the NumPy `maximum` function instead of the `max` function for this purpose:

```
truerange = numpy.maximum(h - l, h - previousclose, previousclose - l)
```

4. **Initializing an atr array**: Create an atr array of size N and initialize its values to 0:

```
atr = numpy.zeros(N)
```

5. **Initializing the first element**: The first value of the array is just the average of the `truerange` array:

```
atr[0] = numpy.mean(truerange)
```

Calculate the other values with the following formula:

$$\frac{((N-1)PATR+TR)}{N}$$

Here, `PATR` is the previous day's ATR; `TR` is the true range:

```
for i in range(1, N):
    atr[i] = (N - 1) * atr[i - 1] + truerange[i]
    atr[i] /= N
```

What just happened?

We formed three arrays, one for each of the three ranges—daily range, the gap between the high of today and the close of yesterday, and the gap between the close of yesterday and the low of today. This tells us how much the stock price moved and, therefore, how volatile it is. The algorithm requires us to find the maximum value for each day. The `max` function that we used before can give us the maximum value within an array, but that is not what we want here. We need the maximum value across arrays, so we want the maximum value of the first elements in the three arrays, the second elements, and so on. In this *Time for action* tutorial, we saw that the `maximum` function can do this. After that, we computed a moving average of the true range values. In the following tutorials, we will learn better ways to calculate moving averages.

Have a go hero – taking the minimum function for a spin

Besides the `maximum` function, there is a `minimum` function. You can probably guess what it does. Make a small script or start an interactive session in IPython to prove your assumptions.

Simple moving average

The **simple moving average** is commonly used to analyze time-series data. To calculate it, we define a moving window of N periods, N days in our case. We move this window along the data and calculate the mean of the values inside the window.

Time for action – computing the simple moving average

The moving average is easy enough to compute with a few loops and the mean function, but NumPy has a better alternative—the convolve function. The simple moving average is, after all, nothing more than a convolution with equal weights or, if you like, unweighted. Use the following steps to compute the simple moving average:

1. **Setting the weights**: Use the ones function to create an array of size N and elements initialized to 1; then, divide the array by N to give us the weights:

```
N = int(sys.argv[1])
weights = numpy.ones(N) / N
print "Weights", weights
```

   For N = 5, this gives us the following output:

```
Weights [ 0.2   0.2   0.2   0.2   0.2]
```

2. **Using the convolve function**: Now call the convolve function with these weights:

```
c = numpy.loadtxt('data.csv', delimiter=',', usecols=(6,),
unpack=True)
sma = numpy.convolve(weights, c)[N-1:-N+1]]
```

3. **Plotting the simple moving average**: From the array that convolve returned, we extracted the data in the center of size N. The following code makes an array of time values and plots with the matplotlib that we will be covering in a later chapter:

```
c = numpy.loadtxt('data.csv', delimiter=',', usecols=(6,),
unpack=True)
sma = numpy.convolve(weights, c)[N-1:-N+1]
t = numpy.arange(N - 1, len(c))
plot(t, c[N-1:], lw=1.0)
plot(t, sma, lw=2.0)
show()
```

In the following chart, the smooth thick line is the 5-day simple moving average and the jagged thin line is the close price:

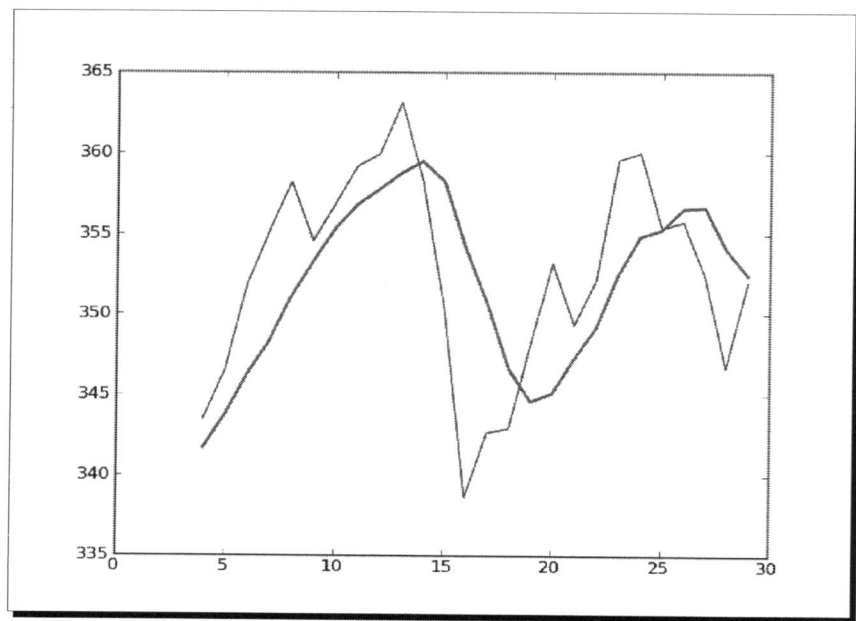

What just happened?

We computed the simple moving average for the close stock price. Truly great riches are within your reach. It turns out that the simple moving average is just a signal processing technique—a convolution with weights 1/N, where N is the size of the moving average window. We learned that the `ones` function can create an array with ones and the `convolve` function calculates the convolution of a data set with specified weights.

Exponential moving average

The exponential moving average is a popular alternative to the simple moving average. This method uses exponentially-decreasing weights. The weights for point in the past decrease exponentially but never reach zero. We will learn about the `exp` and `linspace` function while calculating the weights.

Time for action – calculating the exponential moving average

Given an array, the exp function calculates the exponential of each array element. For example, look at the following code:

```
x = numpy.arange(5)
print "Exp", numpy.exp(x)
```

It gives the following output:

```
Exp [  1.          2.71828183   7.3890561   20.08553692  54.59815003]
```

The linspace function takes, as parameters, a start and a stop and optionally an array size. It returns an array of evenly-spaced numbers. Here is an example:

```
print "Linspace", numpy.linspace(-1, 0, 5)
```

This will give us the following output:

```
Linspace [-1.    -0.75 -0.5  -0.25  0.  ]
```

Let's calculate the exponential moving average for our data:

1. **Initialize the weights**: Now, back to the weights—calculate them with exp and linspace:

   ```
   N = int(sys.argv[1])
   weights = numpy.exp(numpy.linspace(-1., 0., N))
   ```

2. **Normalization**: Normalize the weights. The ndarray object has a sum method that we will use:

   ```
   weights /= weights.sum()
   print "Weights", weights
   ```

   For N = 5, we get these weights:

   ```
   Weights [ 0.11405072  0.14644403  0.18803785  0.24144538
   0.31002201]
   ```

3. **Convolve**: After that, it's easy going—we just use the convolve function that we learned about in the simple moving average tutorial. We will also plot the results:

   ```
   c = numpy.loadtxt('data.csv', delimiter=',', usecols=(6,),
   unpack=True)
   ema = numpy.convolve(weights, c)[N-1:-N+1]
   t = numpy.arange(N - 1, len(c))
   plot(t, c[N-1:], lw=1.0)
   plot(t, ema, lw=2.0)
   show()
   ```

That gives this nice chart where, again, the close price is the thin jagged line and the exponential moving average is the smooth thick line:

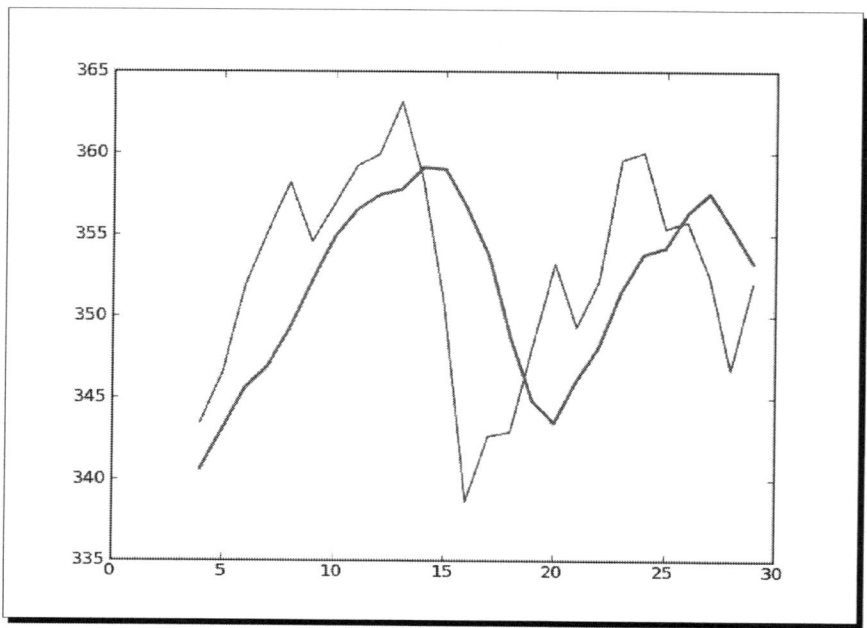

What just happened?

We calculated the exponential moving average of the close price. First, we computed exponentially-decreasing weights with the `exp` and `linspace` functions. `linspace` gave us an array with evenly-spaced elements, and then, we calculated the exponential for these numbers. We called the `ndarray sum` method in order to normalize the weights. After that, we applied the `convolve` trick that we learned in the simple moving average tutorial.

Bollinger bands

Bollinger bands are yet another technical indicator. Yes, there are thousands of them. This one is named after its inventor and consists of three parts: First, a simple moving average. Second, an upper band of two standard deviations above this moving average—the standard deviation is derived from the same data with which the moving average is calculated. Third, a lower band of two standard deviations below the moving average.

Time for action – enveloping with Bollinger bands

We already know how to calculate the simple moving average. So, if you need to, please review the *Time for action* tutorial in this chapter. This example will introduce the NumPy `fill` function. The `fill` function sets the value of an array to a scalar value. The function should be faster than `array.flat = scalar` or setting the values of the array one-by-one in a loop.

1. **Calculate the Bollinger bands**: Starting with an array called `sma` that contains the moving average values, we will loop through all the data sets corresponding to said values. After forming the data set, calculate the standard deviation. Note that it is necessary, at a certain point, to calculate the difference between each data point and the corresponding average value. If we did not have NumPy, we would loop through these points and subtract each of the values one-by-one from the corresponding average. However, the NumPy `fill` function allows us to construct an array having elements set to the same value. This enables us to save on one loop and subtract arrays in one go:

```python
deviation = []
C = len(c)

for i in range(N - 1, C):
    if i + N < C:
        dev = c[i: i + N]
    else:
        dev = c[-N:]:]

    averages = numpy.zeros(N)
    averages.fill(sma[i - N - 1])
    dev = dev - averages
    dev = dev ** 2
    dev = numpy.sqrt(numpy.mean(dev))))
    deviation.append(dev)

deviation = 2 * numpy.array(deviation)
upperBB = sma + deviation
lowerBB = sma - deviation
```

2. **Plot the bands**: To plot, we will use the following code (don't worry about it now; we will see how this works in *Chapter 9, Plotting with Matplotlib*):

```python
t = numpy.arange(N - 1, C)
plot(t, c_slice, lw=1.0)
plot(t, sma, lw=2.0)
plot(t, upperBB, lw=3.0)
plot(t, lowerBB, lw=4.0)
show()
```

Following is a chart of the Bollinger bands for our data. The jagged thin line in the middle represents the close price, the slightly thicker, smoother line crossing it is the moving average:

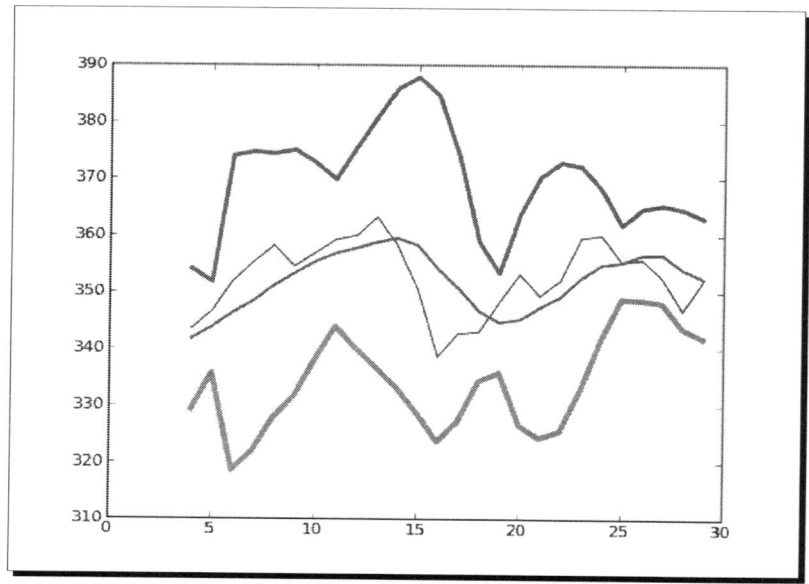

What just happened?

We worked out the Bollinger bands that envelope the close price of our data. More importantly, we got acquainted with the NumPy `fill` function. This function fills an array with a scalar value. This is the only parameter of the `fill` function.

Have a go hero – switching to exponential moving average

It is customary to choose the simple moving average to centre the Bollinger band on. The second-most popular choice is the exponential moving average, so try that as an exercise. You can find a suitable example in this chapter, if you need pointers.

Check that the `fill` function is faster or is as fast as `array.flat = scalar`, or setting the value in a loop.

Linear model

Many phenomena in science have a related linear relationship model. The NumPy `linalg` package deals with linear algebra computations. We will begin with the assumption that a price value can be derived from N previous prices based on a linear relationship relation.

Time for action – predicting price with a linear model

Keeping an open mind, let's assume that we can express a stock price as a linear combination of previous values, that is, a sum of those values multiplied by certain coefficients that we need to determine. In linear algebra terms, this boils down to finding a least-squares solution. The recipe goes as follows.

1. **Form a price vector**: First, form a vector bbx containing N price values:

   ```
   bbx = c[-N:]
   bbx = b[::-1]
   print "bbx", x
   ```

   The result is as follows:

   ```
   bbx [ 351.99  346.67  352.47  355.76  355.36]
   ```

2. **Pre-initialize the matrix**: Second, pre-initialize the matrix A to be N-by-N and contain zeroes:

   ```
   A = numpy.zeros((N, N), float)
   print "Zeros N by N", A

   Zeros N by N [[ 0.  0.  0.  0.  0.]
    [ 0.  0.  0.  0.  0.]
    [ 0.  0.  0.  0.  0.]
    [ 0.  0.  0.  0.  0.]
    [ 0.  0.  0.  0.  0.]]
   ```

3. **Fill the matrix**: Third, fill the matrix A with N preceding price values for each value in bbx:

   ```
   for i in range(N):
      A[i, ] = c[-N - 1 - i: - 1 - i]
   print "A", A
   ```

   Now, A looks like this:

   ```
   A [[ 360.     355.36  355.76  352.47  346.67]
    [ 359.56  360.     355.36  355.76  352.47]
    [ 352.12  359.56  360.     355.36  355.76]
    [ 349.31  352.12  359.56  360.     355.36]
    [ 353.21  349.31  352.12  359.56  360.  ]]
   ```

4. **Get the least squares solution**: The objective is to determine the coefficients that satisfy our linear model, by solving the least-squares problem. Employ the `lstsq` function of the NumPy `linalg` package to do that:

```
(x, residuals, rank, s) = numpy.linalg.lstsq(A, b)

print x, residuals, rank, s
```

The result is as follows:

```
[ 0.78111069 -1.44411737  1.63563225 -0.89905126  0.92009049]
[] 5 [  1.77736601e+03   1.49622969e+01   8.75528492e+00
5.15099261e+00   1.75199608e+00]
```

The tuple returned contains the coefficients xxb that we were after, an array comprising of residuals, the rank of matrix A, and the singular values of A.

5. **Extrapolate to the next day**: Once we have the coefficients of our linear model, we can predict the next price value. GetCompute the dot product (with the NumPy `dot` function) of the coefficients and the last known N prices:

```
print numpy.dot(b, x)
```

The dot product is the linear combination of the coefficients xxb and the prices x. As a result, we get:

```
357.939161015
```

I looked it up; the actual close price of the next day was 353.56. So, our estimate with N = 5 was not that far off.

What just happened?

We predicted tomorrow's stock price today. If this works in practice, we could retire early! See, this book was a good investment after all! We designed a linear model for the predictions. The financial problem was reduced to a linear algebraic one. NumPy's `linalg` package has a practical `lstsq` function that helped us with the task at hand—estimating the coefficients of a linear model. After obtaining a solution, we plugged the numbers in the NumPy `dot` function that presented us an estimate through linear regression.

Trend lines

A trend line is a line among a number of so-called pivot points on a stock chart. As the name suggests, the line's trend portrays the trend of the price development. In the past, traders drew trend lines on paper; but, nowadays, we can let a computer draw it for us. In this tutorial, we shall resort to a very simple approach that is probably not very useful in real life, but it should clarify the principle well.

Time for action – drawing trend lines

Follow the ensuing steps to draw trend lines:

1. **Determine the pivots**: First, we need to determine the pivot points. We shall pretend they are equal to the arithmetic mean of the high, low, and close price:

```
h, l, c = numpy.loadtxt('data.csv', delimiter=',', usecols=(4, 5, 6), unpack=True)

pivots = (h + l + c) / 3
print "Pivots", pivots
```

From the pivots, we can deduce the so-called resistance and support levels. The support level is the lowest level at which the price rebounds. The resistance level is the highest level at which the price bounces back. These are not natural phenomena, mind you, they are merely estimates. Based on these estimates, it is possible to draw support and resistance trend lines. We will define the daily spread to be the difference of the high and low price.

2. **Fit data to a line**: Define a function to fit line to data to a line where $y = at + b$. The function should return a and b. This is another opportunity to apply the `lstsq` function of the NumPy `linalg` package. Rewrite the line equation to $y = Ax$, where $A = [t\ 1]$ and $x = [a\ b]$. Form A with the NumPy `ones` and `vstack` function:

```
def fit_line(t, y):
    A = numpy.vstack([t, numpy.ones_like(t)]).))]).T
    return numpy.linalg.lstsq(A, y)[0]
```

3. **Determine the support and resistance levels**: Assuming that support levels are one daily spread below the pivots, and that resistance levels are one daily spread above the pivots, fit the support and resistance trend lines:

```
t = numpy.arange(len(c))
sa, sb = fit_line(t, pivots - (h - l))
ra, rb = fit_line(t, pivots + (h - l))
support = sa * t + sb
resistance = ra * t + rb
```

4. **Analyze the bands**: At this juncture, we have all the necessary information to draw the trend lines, however, it is wise to check how many points fall between the support and resistance levels. Obviously, if only a small percentage of the data is between the trend lines, then this setup is of no use to us. Make up a condition for points between the bands and select with the `where` function based on the condition:

```
condition = (c > support) & (c < resistance)
print "Condition", condition
between_bands = numpy.where(condition)
```

These are the condition values:

```
Condition [False False  True  True  True  True  True False False
True False False
  False False False  True False False False  True  True  True  True
False False  True  True  True False  True]
```

Double-check the values:

```
print support[between_bands]
print c[between_bands]
print resistance[between_bands]
```

The array returned by the `where` function has rank 2, so call the `ravel` function before calling the `len` function:

```
between_bands = len(numpy.ravel(between_bands))))
print "Number points between bands", between_bands
print "Ratio between bands", float(between_bands)/len(c)
```

You will get the following result:

```
Number points between bands 15
Ratio between bands 0.5
```

As an extra bonus, we gained a predictive model. Extrapolate the next day resistance and support levels:

```
print "Tomorrows support", sa * (t[-1] + 1) + sb
print "Tomorrows resistance", ra * (t[-1] + 1) + rb
```

This results in:

```
Tomorrows support 349.389157088
Tomorrows resistance 360.749340996
```

Another approach to figure out how many points are between the support and resistance estimates is to use `[]` and `intersect1d`. Define selection criteria in the `[]` operatpr and intersect the results with the intersect1d function.

```
a1 = c[c > support]
a2 = c[c < resistance]
print "Number of points between bands 2nd approach" ,len(numpy.
intersect1d(a1, a2))
```

Not surprisingly, we get:

```
Number of points between bands 2nd approach 15
```

5. **Plot the bands**: Once more, we will plot the results:

```
plot(t, c)
plot(t, support)
plot(t, resistance)
show()
```

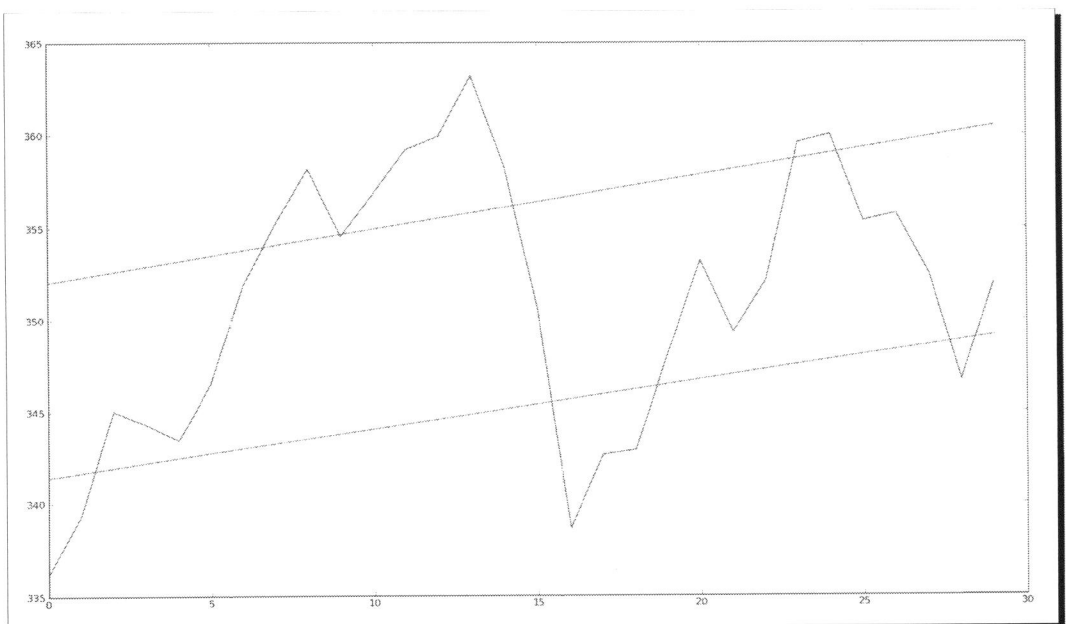

In the preceding plot, we have the price data and the corresponding support and resistance lines.

What just happened?

We drew trend lines without having to mess around with rulers, pencils, and paper charts. We defined a function that can fit data to a line with the NumPy `vstack`, `ones`, and `lstsq` functions. We fit the data in order to define support and resistance trend lines. Then we figured out how many points are within the support and resistance range. We did this using two separate methods that produced the same result.

The first method used the `where` function with a Boolean condition. The second method made use of the `[]` operator and the `intersect1d` function. The `intersect1d` function returns an array of common elements from two arrays.

Methods of ndarray

The NumPy ndarray class has a lot of methods that work on the array. Most of the time, these methods return an array. You may have noticed that many of the functions that are part of the NumPy library have a counterpart with the same name and functionality in the ndarray object. This is mostly due to the historical development of NumPy.

The list of ndarray methods is pretty long, so we cannot cover them all. The mean, var, sum, std, argmax, argmin, and mean functions that we saw earlier are also ndarray methods.

To clip and compress arrays, look at the following section:

Time for action – clipping and compressing arrays

1. Here are a few examples of ndarray methods. The clip method returns a clipped array, so that all values above a maximum value are set to the maximum and values below a minimum are set to the minimum value. Clip an array with values 0 to 4 to 1 and 2:

```
a = numpy.arange(5)
print "a =", a
print "Clipped", a.clip(1, 2)
```

This gives the following output:

```
a = [0 1 2 3 4]
Clipped [1 1 2 2 2]
```

2. The ndarray compress method returns an array based on a condition. For instance, look at the following code:

```
a = numpy.arange(4)
print a
print "Compressed", a.compress(a > 2)
```

This returns the following output:

```
[0 1 2 3]
Compressed [3]
```

What just happened?

We created an array with values 0 to 3 and selected the last element with the `compress` function based on the condition `a > 2`.

Factorial

Many programming books have an example of calculating the factorial. We should not break with this tradition.

Time for action – calculating the factorial

The `ndarray` has the `prod` method, which computes the product of the elements in an array.

1. **Call the prod function:** Calculate the factorial of eight. To do that, generate an array with values 1 to 8 and call the `prod` function on it:

   ```
   b = numpy.arange(1, 9)
   print "b =", b
   print "Factorial", b.prod()
   ```

   Check the result with your pocket calculator:

   ```
   b = [1 2 3 4 5 6 7 8]
   Factorial 40320
   ```

   This is nice, but what if we want to know all the factorials from 1 to 8?

2. **Call cumprod:** No problem! Call the `cumprod` method, which computes the cumulative product of an array:

   ```
   print "Factorials", b.cumprod()
   ```

   It's pocket calculator time again:

   ```
   Factorials [    1     2     6    24   120   720  5040 40320]
   ```

What just happened?

We used the `prod` and `cumprod` functions to calculate factorials.

Summary

This chapter informed us about a great number of common NumPy functions. We read a file with `loadtxt` and wrote to a file with `savetxt`. We made an identity matrix with the `eye` function. We read a CSV file containing stock quotes with the `loadtxt` function. The NumPy `average` and `mean` functions allow one to calculate the weighed average and arithmetic mean of a data set.

A few common statistics functions were also mentioned: First, the `min` and `max` functions we used to determine the range of the stock prices. Second, the `median` function that gives the median of a data set. Finally, the `std` and `var` functions that return the standard deviation and variance of a set of numbers.

We calculated the simple stock returns with the `diff` function that returns the back differences between sequential elements. The `log` function computes the natural logarithms of array elements.

By default, `loadtxt` tries to convert all data into floats. The `loadtxt` function has a special parameter for this purpose. The parameter is called `converters` and is a dictionary that links columns with the so-called converter functions.

We defined a function and passed it as an argument to the `apply_along_axis` function. We implemented an algorithm with the requirement to find the maximum value across arrays.

We learned that the `ones` function can create an array with ones and the `convolve` function calculates the convolution of a data set with specified weights.

We computed exponentially-decreasing weights with the `exp` and `linspace` functions. `Linspace` gave us an array with evenly-spaced elements, and then we calculated the exponential for these numbers. We called the `ndarray` `sum` method in order to normalize the weights.

We got acquainted with the NumPy `fill` function. This function fills an array with a scalar value, the only parameter of the `fill` function.

After this tour through the common NumPy functions, we will continue covering convenience NumPy functions in the next chapter.

4
Convenience Functions for Your Convenience

As we have noticed, NumPy has a great number of functions. Many of these functions are there just for your convenience. Knowing these functions will greatly increase your productivity. This includes functions that select certain parts of your arrays (based on a Boolean condition, for instance) or manipulate polynomials. An example of computing correlation of stock returns is given to give you a taste of data analysis in NumPy.

In this chapter, we shall cover the following topics:

- ◆ Data selection and extraction
- ◆ Simple data analysis
- ◆ Examples of correlation of returns
- ◆ Polynomials
- ◆ Linear algebra functions

In the previous chapter, we had one single data file to play around with. Things have significantly improved in this chapter—we now have two data files. Let's go ahead and explore the data with NumPy.

Correlation

Have you noticed that the stock price of some companies is closely followed by another one, usually a rival in the same sector? The theoretical explanation is that, because these two companies are in the same type of business, they share the same challenges, require the same materials and resources, and compete for the same type of customers.

You could think of many possible pairs, but you would want to check whether a real relationship exists. One way is to have a look at the correlation of the stock returns of both stocks. A high correlation implies a relationship of some sort. It is not proof though, especially if you don't use sufficient data.

Time for action – trading correlated pairs

For this tutorial, we will use two sample data sets, containing the bare minimum of end-of-day price data. The first company is **BHP Billiton (BHP)**, which is active in mining of petroleum, metals, and diamonds. The second is **Vale (VALE)**, which is also a metals and mining company. So there is some overlap, albeit not hundred percent. For trading correlated pairs, follow the ensuing steps:

1. **Load the data**: First, load the data, specifically the close price of the two securities, from the CSV files in the example code directory of this chapter and calculate the returns. If you don't remember how to do it, there are plenty of examples in the previous chapter.

2. **Covariance**: Covariance tells us how two variables vary together; it is nothing more than unnormalized correlation. Compute the covariance matrix from the returns with the cov function (it's not strictly necessary to do this, but it will allow us to demonstrate a few matrix operations):

```
covariance = numpy.cov(bhp_returns, vale_returns)
print "Covariance", covariance
```

The covariance matrix is as follows:

```
Covariance [[ 0.00028179  0.00019766]
 [ 0.00019766  0.00030123]]
```

3. **Diagonal values**: View the values on the diagonal with the diagonal function:

```
print "Covariance diagonal", covariance.diagonal()
```

The diagonal values of the covariance matrix are as follows:

```
Covariance diagonal [ 0.00028179  0.00030123]
```

Notice that the values on the diagonal are not equal to each other, this is different from the correlation matrix.

4. **Trace**: Compute the trace, the sum of the diagonal values, with the `trace` function:

```
print "Covariance trace", covariance.trace()
```

The trace values of the covariance matrix are as follows:

```
Covariance trace 0.00058302354992
```

5. **Correlation from covariance**: The correlation of two vectors is defined as the covariance, divided by the product of the respective standard deviations of the vectors. The equation for vectors a and b is:

$$Corr(a,b) = \frac{cov(a,b)}{\sigma_b \sigma_b}$$

Try it out:

```
print covariance/ (bhp_returns.std() * vale_returns.std())
```

The correlation matrix is as follows:

```
[[ 1.00173366   0.70264666]
 [ 0.70264666   1.0708476 ]]
```

6. **Correlation coefficients**: We will measure the correlation of our pair with the correlation coefficient. The correlation coefficient takes values between -1 to 1. The correlation of a set of values with itself is 1 by definition. This would be the ideal value; however, we will be also happy with a slightly lower value. Calculate the correlation coefficient (or, more accurately, the correlation matrix) with the `corrcoef` function:

```
print "Correlation coefficient", numpy.corrcoef(bhp_returns, vale_returns)
```

The coefficients are as follows:

```
[[ 1.            0.67841747]
 [ 0.67841747   1.        ]]
```

The values on the diagonal are just the correlations of the BHP and VALE with themselves and are, therefore, equal to 1. In all probability, no real calculation takes place. The other two values are equal to each other since correlation is symmetrical, meaning that the correlation of BHP with VALE is equal to the correlation of VALE with BHP. It seems that the correlation is not that strong.

7. **Breakout**: Another important point is whether the two stocks under consideration are in sync or not. Two stocks are considered out of sync if their difference is two standard deviations from the mean of the differences.

If they are out of sync, we could initiate a trade, hoping that they eventually will get back in sync again. Compute the difference between the close prices of the two securities to check the synchronization:

```
difference = bhp - vale
```

Check whether the last difference in price is out of sync; see the following code:

```
avg = numpy.mean(difference)
dev = numpy.std(difference)
print "Out of sync", numpy.abs(difference[-1] - avg) > 2 * dev
```

Unfortunately, we cannot trade yet:

```
Out of sync False
```

8. **Plotting**: Plotting requires `Matplotlib`; this will be discussed in *Chapter 9, Plotting with Matplotlib*. Plotting can be done as follows:

```
t = numpy.arange(len(bhp_returns))
plot(t, bhp_returns, lw=1)
plot(t, vale_returns, lw=2)
show()
```

The resulting plot:

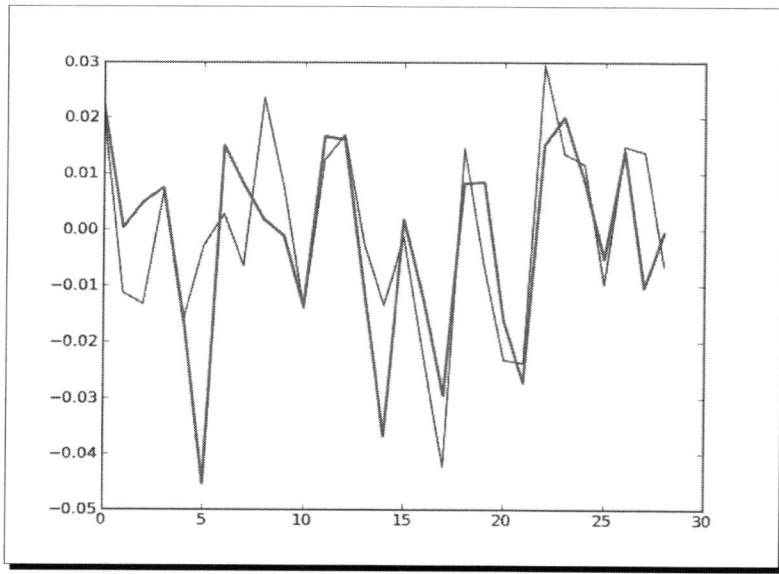

What just happened?

We analyzed the relation of the closing stock prices of BHP and VALE. To be precise, we calculated the correlation of their stock returns. This was achieved with the `corrcoef` function. Further, we saw how the covariance matrix can be computed, from which the correlation can be derived. As a bonus, a demonstration was given of the `diagonal` and `trace` functions that can give us the diagonal values and the trace of a matrix, respectively.

Pop quiz – calculating covariance

1. Which function returns the covariance of two arrays?

    a. covariance

    b. covar

    c. cov

    d. cvar

Polynomials

Do you like calculus? Me, I love it! One of the ideas in calculus is Taylor expansion, that is, representing a differentiable function as an infinite series. In practice, this means that any differentiable, and therefore continuous, function can be estimated by a polynomial of a high degree. The terms of the higher degree would then be assumed to be negligibly small.

Time for action – fitting to polynomials

The NumPy `polyfit` function can fit a set of data points to a polynomial even if the underlying function is not continuous.

1. **Polynomial fit**: Continuing with the price data of BHP and VALE, let's look at the difference of their close prices and fit it to a polynomial of the third power:

```
bhp=numpy.loadtxt('BHP.csv', delimiter=',', usecols=(6,),
   unpack=True)
vale=numpy.loadtxt('VALE.csv', delimiter=',', usecols=(6,),
   unpack=True)
t = numpy.arange(len(bhp))
poly = numpy.polyfit(t, bhp - vale, int(sys.argv[1]))
print "Polynomial fit", poly
```

The polynomial fit (in this example, a cubic polynomial was chosen):

```
Polynomial fit [  1.11655581e-03  -5.28581762e-02   5.80684638e-01
5.79791202e+01]
```

2. Extrapolate to the next day: The numbers you see are the coefficients of the polynomial. Extrapolate to the next value with the `polyval` function and the polynomial object we got from the fit:

```
print "Next value", numpy.polyval(poly, t[-1] + 1)
```

The next value we predict will be:

```
Next value 57.9743076081
```

3. Find the roots: Ideally, the difference between the close prices of BHP and VALE should be as small as possible. In an extreme case, it might be zero at some point. Find out when our polynomial fit reaches zero with the `roots` function:

```
print "Roots", numpy.roots(poly)
```

The roots of the polynomial are as as follows:

```
Roots [ 35.48624287+30.62717062j   35.48624287-30.62717062j
 -23.63210575 +0.j         ]
```

The roots are complex; that's no good.

4. Differentiate: Another thing we learned in calculus class was to find **extremums**—these could be potential maxima or minima. Remember, from calculus, that these are the points where the derivative of our function is zero. Differentiate the polynomial fit with the `polyder` function:

```
der = numpy.polyder(poly)
print "Derivative", der
```

The coefficients of the derivative polynomial are as follows:

```
Derivative [ 0.00334967 -0.10571635  0.58068464]
```

The numbers you see are the coefficients of the derivative polynomial.

5. Find the extrema: Get the roots of the derivative:

```
print "Extremas", numpy.roots(der)
```

The extremas that we get are:

```
Extremas [ 24.47820054   7.08205278]
```

Let's double check; compute the values of the fit with `polyval`:

```
vals = numpy.polyval(poly, t)
```

6. Double-check: Now, find the maximum and minimum values with `argmax` and `argmin`:

```
vals = numpy.polyval(poly, t)
print numpy.argmax(vals)
print numpy.argmin(vals)
```

This gives us the expected results. Ok, not quite the same results, but, if we backtrack to step 1, we can see that `t` was defined with the `arange` function:

```
7
24
```

7. **Plot**: Plot the data and the fit it as follows:

```
plot(t, bhp - vale)
plot(t, vals)
show()
```

It results in this plot:

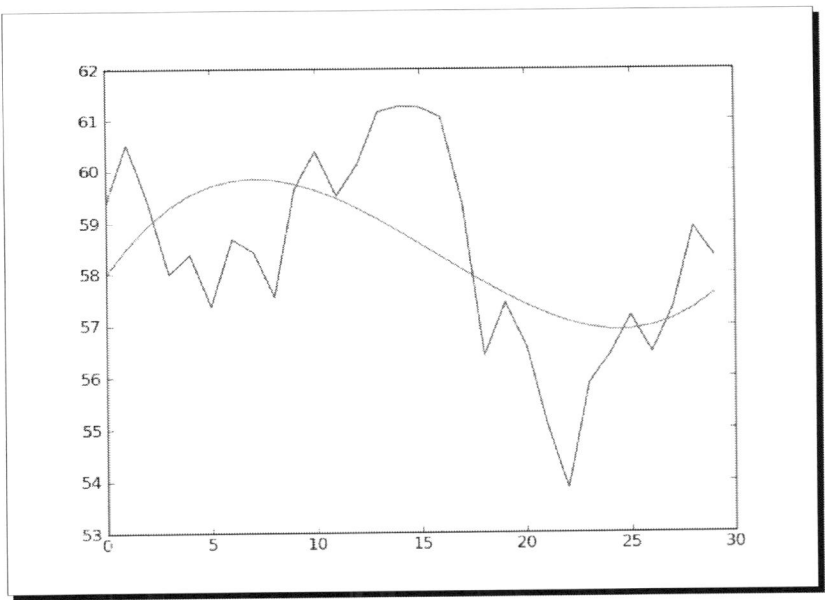

Obviously, the smooth line is the fit and the jagged line is the underlying data. It's not that good a fit, so you might want to try a higher order polynomial.

What just happened?

We fit data to a polynomial with the `polyfit` function. We learned about the `polyval` function that computes the values of a polynomial, the `roots` function that returns the roots of the polynomial, and the `polyder` function that gives back the derivative of a polynomial.

Have a go hero – improving the fit

There are a number of things you could do to improve the fit. Try a different power as, in this tutorial, a cubic polynomial was chosen. Consider smoothing the data before fitting it. One way you could smooth is with a moving average. Examples of simple and exponential moving average calculations can be found in the previous chapter.

On-balance volume

Volume is a very important variable in investing; it indicates how big a price move is. The on-balance volume indicator is one of the simplest stock price indicators. It is based on the close price of the current and previous days and the volume of the current day. For each day, if the close price today is higher than the close price of yesterday then the value of the on-balance volume is equal to the volume of today. On the other hand, if today's close price is lower than yesterday's close price then the value of the on-balance volume indicator is the difference between the on-balance volume and the volume of today. If the close price did not change then the value of the on-balance volume is zero.

Time for action – balancing volume

In other words we need to multiply the sign of the close price with the volume. In this tutorial, we will go over two approaches to this problem, one using the NumPy `sign` function, and the other using the NumPy `piecewise` function.

1. **Load the data**: Load the BHP data into a close and volume array:

   ```
   c, v=numpy.loadtxt('BHP.csv', delimiter=',', usecols=(6, 7),
     unpack=True)
   ```

   Compute the absolute value changes. Calculate the change of the close price with the `diff` function. The `diff` function computes the difference between two sequential array elements and returns an array containing these differences:

   ```
   change = numpy.diff(c)
   print "Change", change
   ```

   The changes of the close price are shown as follows:

   ```
   Change [ 1.92 -1.08 -1.26  0.63 -1.54 -0.28  0.25 -0.6   2.15
   0.69 -1.33  1.16
     1.59 -0.26 -1.29 -0.13 -2.12 -3.91  1.28 -0.57 -2.07 -2.07  2.5
   1.18
   -0.88  1.31  1.24 -0.59]
   ```

2. **Determine the signs**: The NumPy `sign` function returns the signs for each element in an array. `-1` is returned for a negative number, `1` for a positive number, and `0`, otherwise. Apply the `sign` function to the `change` array:

```
signs = numpy.sign(change)
print "Signs", signs
```

The signs of the change array are as follows:

```
Signs [ 1. -1. -1.  1. -1. -1.  1. -1.  1.  1. -1.  1.  1. -1. -1.
 -1. -1. -1.
 -1. -1. -1.  1.  1.  1. -1.  1.  1. -1.]
```

Alternatively, we can calculate the signs with the `piecewise` function. The `piecewise` function, as its name suggests, evaluates a function piece-by-piece. Call the function with the appropriate return values and conditions:

```
pieces = numpy.piecewise(change, [change < 0, change > 0], [-1,
  1])
print "Pieces", pieces
```

The signs are shown again, as follows:

```
Pieces [ 1. -1. -1.  1. -1. -1.  1. -1.  1.  1. -1.  1.  1. -1.
 -1. -1. -1. -1.
 -1. -1. -1.  1.  1.  1. -1.  1.  1. -1.]
```

Check that the outcome is the same:

```
print "Arrays equal?", numpy.array_equal(signs, pieces)
```

And the outcome is as follows:

```
Arrays equal? True
```

3. **On-balance volume**: The on-balance volume depends on the change of the previous close, so we can not calculate it for the first day in our sample:

```
print "On balance volume", v[1:] * signs
```

The on-balance volume is as follows:

```
[ 2620800. -2461300. -3270900.  2650200. -4667300. -5359800.
7768400.
 -4799100.  3448300.  4719800. -3898900.  3727700.  3379400.
-2463900.
 -3590900. -3805000. -3271700. -5507800.  2996800. -3434800.
-5008300.
 -7809799.  3947100.  3809700.  3098200. -3500200.  4285600.
3918800.
 -3632200.]
```

What just happened?

We computed the on-balance volume that depends on the change of the closing price. Using the NumPy `sign` and `piecewise` functions, we went over two different methods to determine the sign of the change.

The mode

In statistics, the mode summarizes a set of values, just like an average or median. The mode is the most frequent value or values. For instance, if we have the values 0, 1, 2, 2, 3, then the mode would be 2. The mode doesn't have to be a unique number. The mode could consist of multiple numbers as long as these numbers are the most frequent ones. For example, if we have the numbers 1, 1, 2, 2, 3, the mode would be 1 and 2.

Time for action – determining the mode of stock returns

When it comes to determining the mode of stock returns we can do two things—we can try to find the peak of the histogram of said returns, or we can turn the returns into integers and find the mode of those integers. If we don't do this, it is impossible to determine the mode; we would have to deal with a large number of unique numbers.

1. **Unique numbers**: The `unique` function returns the unique numbers of an array. Here is an example of how it works:

   ```
   print "Unique", numpy.unique(numpy.array([2, 2]))
   ```

   As you would expect, one unique number is returned:

   ```
   Unique [2]
   ```

   Now, load the data and find out how many unique stock returns there are:

   ```
   bhp = numpy.loadtxt('BHP.csv', delimiter=',', usecols=(6,),
     unpack=True)
   bhp_returns = numpy.diff(bhp) / bhp[ : -1]
   print "BHP returns", bhp_returns
   print "Total number", len(bhp_returns), "Unique number",
     len(numpy.unique(bhp_returns))
   ```

   Just as suspected, there are a lot of unique stock returns; actually, all of them are unique.

   ```
   BHP returns [ 0.02048656 -0.01129235 -0.01332487  0.00675241
   -0.01639519 -0.00303063
      0.00271415 -0.00649632  0.02343069  0.00734746 -0.0140592
   0.01243701
   ```

```
    0.01683787 -0.00270777 -0.01347118 -0.0013761   -0.02247191
-0.04239861
    0.01449439 -0.00636232 -0.0232532  -0.02380679  0.02945335
0.01350423
    0.01163053 -0.00982252  0.01476722  0.01377472 -0.00646504]
Total number 29 Unique number 29
```

2. **Histogram**: NumPy has a `histogram` function that we will use. This method is sensitive to the number of bins. Set the number of bins to the square root of the number of array elements (this is a rule of thumb that works quite well):

```
nbins = numpy.sqrt(len(bhp_returns))
```

Call the `histogram` function with the number of bins we calculated. It's shown as follows:

```
N, bins = numpy.histogram(bhp_returns, bins=nbins)
print "Counts", N, "Bins", bins
```

The `histogram` function returns the number of occurrences within each bin and the bins themselves:

```
Counts [ 1  5 10  9  4] Bins [-0.04239861 -0.02802822 -0.01365783
0.00071256  0.01508295  0.02945335]
```

Determine the mode by finding the bin corresponding to the highest count in the histogram:

```
index_max = N.argmax()
print "mode", bins[index_max]
```

The mode is as follows:

```
mode -0.0136578288488
```

3. **Converting to promilles**: It is a bit arbitrary whether we should convert the stock returns to `promilles` or percentages, as we have integer values. Convert the stock returns to `promilles` with the `astype` function:

```
bhp_promilles = (bhp_returns * 1000).astype(int)
```

4. **Sort**: Sort the values by calling the `sort` function:

```
sorted = numpy.sort(bhp_promilles)
print "Sorted", sorted
```

The sorted values should be as follows:

```
Sorted [-42 -23 -23 -22 -16 -14 -13 -13 -11  -9  -6  -6  -6  -3
-2  -1   2   6
   7  11  12  13  13  14  14  16  20  23  29]
```

5. **Indices of changed values**: Now, we need to find the indices where values changed:

```
diffed = numpy.diff(sorted)
#values changed
indices = numpy.where(diffed > 0)
print "Indices where values changed", indices
```

The indices are as follows:

```
Indices where values changed (array([ 0,   2,   3,   4,   5,   7,   8,
9, 12, 13, 14, 15, 16, 17, 18, 19, 20,
       22, 24, 25, 26, 27]),)
```

6. **Number of repeats**: Figure out the number of repeats:

```
# number of repeats
repeats = numpy.diff(indices)
print "Repeats", repeats
```

The repeats that we get are:

```
Repeats [[2 1 1 2 1 1 3 1 1 1 1 1 1 1 2 2 1 1 1]]
```

7. **Index with the most repeats**: Find the index with the most repeats:

```
most_repeats_index = numpy.argmax(repeats)
print "Most repeats index", most_repeats_index
```

The index with the most repeats is:

```
Most repeats index 7
```

8. **Index in the sorted array**: Locate the index in the original sorted array of the most frequent `promille` value:

```
index = numpy.ravel(indices)[most_repeats_index + 1]
print "Index", index
```

The index with the most frequent promille value is:

```
Index 12
```

9. **The mode from the sorted array**: Get the mode. Look at the following code:

```
print "Mode", sorted[index]
```

The mode of the array is:

```
Mode -6
```

What just happened?

We determined the mode of sample values with a histogram and by converting the values to integers. The most important thing is that we learned about the `histogram` function.

Simulation

Often, you would want to try something out. Play around, experiment, but preferably without blowing things up or getting dirty. NumPy is perfect for experimentation. We will use NumPy to simulate a trading day, without actually losing money. Many people like to buy on the dip or, in other words, wait for the price of stocks to drop before buying. A variant of that is to wait for the price to drop a small percentage say 0.1 percent below the opening price of the day.

Time for action – avoiding loops with vectorize

The `vectorize` function is a yet another trick to reduce the number of loops in your programs. We will let it calculate the profit of a single trading day:

1. Load data: First, load the data:

```
o, h, l, c = numpy.loadtxt('BHP.csv', delimiter=',', usecols=(3,
4, 5, 6), unpack=True)
```

2. Call vectorize: The `vectorize` function is the NumPy equivalent of the Python `map` function. Call the `vectorize` function, giving it as an argument the `calc_profit` function that we still have to write:

```
func = numpy.vectorize(calc_profit)
```

3. Apply func: We can now apply `func` as if it is a function. Apply the `func` result that we got, to the price arrays:

```
profits = func(o, h, l, c)
```

4. Write the function: The `calc_profit` function is pretty simple. First, we try to buy slightly below the open price. If this is outside of the daily range, then, obviously, our attempt failed and no profit was made, or we incurred a loss, therefore we will return `0`. Otherwise, we sell at the close price and the profit is just the difference between the buy price and the close price. Actually, it is more interesting to have a look at the relative profit:

```
def calc_profit((open, high, low, close):
    #buy just below the open
    buy = open * float(sys.argv[1])
    # daily range
```

```
    if low <  buy < high:
        return (close - buy)/buy
    else:
        return 0
print "Profits", profits
```

5. **Summary of the trades**: There are two days with zero profits: there was either no net gain, or a loss. Select the days with trades and calculate averages:

```
real_trades = profits[profits != 0]
print "Number of trades", len(real_trades), round(100.0 *
len(real_trades)/len(c), 2), "%"
print "Average profit/loss %", round(numpy.mean(real_trades) *
100, 2)
```

The trades summary are shown as follows:

```
Number of trades 28 93.33 %
Average profit/loss % 0.02
```

6. **Winning trades**: As optimists, we are interested in winning trades with a gain greater than zero. Select the days with winning trades and calculate averages:

```
winning_trades = profits[profits > 0]
print "Number of winning trades", len(winning_trades), round(100.0
 * len(winning_trades)/len(c), 2), "%"
print "Average profit %", round(numpy.mean(winning_trades) * 100,
 2)
```

The winning trades are:

```
Number of winning trades 16 53.33 %
Average profit % 0.72
```

7. **Losing trades**: As pessimists, we are interested in losing trades with profit less than zero. Select the days with losing trades and calculate averages:

```
losing_trades = profits[profits < 0]
print "Number of losing trades", len(losing_trades), round(100.0 *
 len(losing_trades)/len(c), 2), "%"
print "Average loss %", round(numpy.mean(losing_trades) * 100, 2)
```

The losing trades are:

```
Number of losing trades 12 40.0 %
Average loss % -0.92
```

What just happened?

We vectorized a function, which is just another way to avoid using loops. We simulated a trading day with a function, which returned the relative profit of each day's trade. We printed a summary of the losing and winning trades.

Have a go hero – analyzing consecutive wins and losses

Although the average profit is positive, it is also important to know whether we had to endure a long streak of consecutive losses. If this is the case, we might be left with little or no capital, and then the average profit would not matter that much.

Find out if there was such a losing streak. If you want, you can also find out if there was a prolonged winning streak.

Smoothing

Noisy data is difficult to deal with, so we often need to do some smoothing. Besides calculating moving averages, we can use one of the NumPy functions to smooth data.

The hanning function is a windowing function formed by a weighted cosine. There are other window functions that will be covered in greater detail in later chapters.

Time for action – smoothing with the hanning function

We will use the hanning function to smooth arrays of stock returns, as shown in the following steps:

1. **Computing the weights**: Call the hanning function to compute weights, for a certain N length window (in this example, N is 8):

   ```
   N = int(sys.argv[1])
   weights = numpy.hanning(N)
   print "Weights", weights
   ```

   The weights are as follows:

   ```
   Weights [ 0.          0.1882551   0.61126047  0.95048443
   0.95048443  0.61126047
     0.1882551   0.          ]
   ```

2. **Smoothing the stock returns**: Calculate the stock returns for the BHP and VALE quotes using `convolve` with normalized `weights`:

```
bhp = numpy.loadtxt('BHP.csv', delimiter=',', usecols=(6,),
    unpack=True)
bhp_returns = numpy.diff(bhp) / bhp[ : -1]
smooth_bhp = numpy.convolve(weights/weights.sum(), bhp_returns)
    [N-1:-N+1]
vale = numpy.loadtxt('VALE.csv', delimiter=',', usecols=(6,),
    unpack=True)
vale_returns = numpy.diff(vale) / vale[ : -1]
smooth_vale = numpy.convolve(weights/weights.sum(), vale_returns)
    [N-1:-N+1]
```

3. **Plotting**: Plotting with `Matplotlib`:

```
t = numpy.arange(N - 1, len(bhp_returns))
plot(t, bhp_returns[N-1:], lw=1.0)
plot(t, smooth_bhp, lw=2.0)
plot(t, vale_returns[N-1:], lw=1.0)
plot(t, smooth_vale, lw=2.0)
show()
```

The chart would appear as follows:

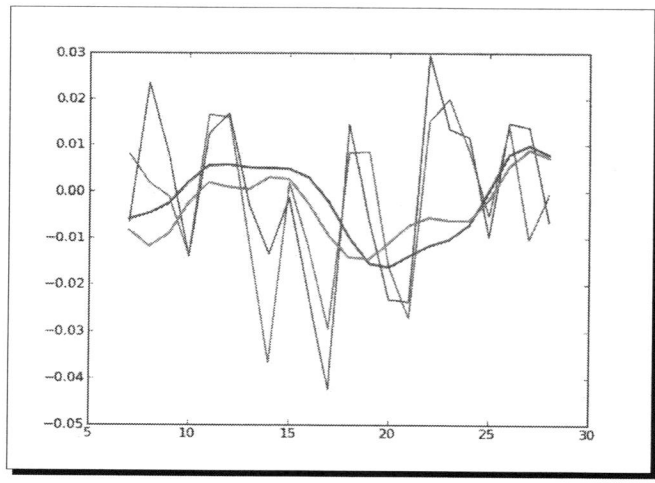

The thin lines on the chart are the stock returns and the thick lines are the result of smoothing. As you can see, the lines cross a few times. These points might be important, because the trend might have changed there. Or, at least, the relation of BHP to VALE might have changed. These turning inflection points probably occur often, so we might want to project into the future.

4. **Fitting to polynomials**: Fit the result of the smoothing step to polynomials:

```
K = int(sys.argv[1])
t = numpy.arange(N - 1, len(bhp_returns))
poly_bhp = numpy.polyfit(t, smooth_bhp, K)
poly_vale = numpy.polyfit(t, smooth_vale, K)
```

5. **Find the crossing points**: Now, we need to compute for a situation where the polynomials we found in the previous step are equal to each other. This boils down to subtracting the polynomials and finding the roots of the resulting polynomial. Subtract the polynomials using `polysub`:

```
poly_sub = numpy.polysub(poly_bhp, poly_vale)
xpoints = numpy.roots(poly_sub)
print "Intersection points", xpoints
```

The points are shown as follows:

```
Intersection points [ 27.73321597+0.j          27.51284094+0.j
24.32064343+0.j
   18.86423973+0.j        12.43797190+1.73218179j   12.43797190-
1.73218179j
    6.34613053+0.62519463j    6.34613053-0.62519463j]
```

6. **Getting the real numbers**: The numbers we get are complex; that is not good for us, unless there is such a thing as imaginary time. Check which numbers are real with the `isreal` function:

```
reals = numpy.isreal(xpoints)
print "Real number?", reals
```

The result is as follows:

```
Real number? [ True   True   True   True False False False False]
```

Some of the numbers are real, so select them with the `select` function. The `select` function forms an array by taking elements from a list of choices, based on a list of conditions:

```
xpoints = numpy.select([reals], [xpoints])
xpoints = xpoints.real
print "Real intersection points", xpoints
```

The real intersection points are as follows:

```
Real intersection points [ 27.73321597  27.51284094  24.32064343
18.86423973   0.         0.   0.   0.]
```

7. **Stripping zeroes**: We managed to pick up some zeroes. The `trim_zeros` function strips the leading and trailing zeros from a one-dimensional array. Get rid of the zeroes with `trim_zeros`:

```
print "Sans 0s", numpy.trim_zeros(xpoints)
```

The zeroes are gone, and the output is shown as follows:

```
Sans 0s [ 27.73321597  27.51284094  24.32064343  18.86423973]
```

What just happened?

We applied the `hanning` function to smooth arrays containing stock returns. We subtracted two polynomials with the `polysub` function. We checked for real numbers with the `isreal` function and selected the real numbers with the `select` function. Finally, we stripped zeroes from an array with the `strip_zeroes` function.

Have a go hero – smoothing variations

Experiment with the other smoothing functions—`hamming`, `blackman`, `bartlett`, and `kaiser`. They work more-or-less in the same way as `hanning`.

Summary

We calculated the correlation of the stock returns of two stocks with the `corrcoef` function. As a bonus, a demonstration of the `diagonal` and `trace` functions was given, which can give us the diagonal and trace of a matrix.

We fit data to a polynomial with the `polyfit` function. We learned about the `polyval` function that computes the values of a polynomial, the `roots` function that returns the roots of the polynomial, and the `polyder` function that gives back the derivative of a polynomial.

Hopefully, we increased our productivity so that we can continue in the next chapter with matrices and universal functions (ufuncs).

5
Working with Matrices and ufuncs

This chapter covers **matrices** *and* **universal functions (ufuncs)**. *Matrices are well known in mathematics and have their representation in NumPy as well. Universal functions work on arrays, element-by-element, or on scalars. ufuncs expect a set of scalars as input and produce a set of scalars as output. Universal functions can typically be mapped to mathematical counterparts, such as, add, subtract, divide, multiply, and so on. We will also be introduced to trigonometric, bitwise, and comparison universal functions.*

In this chapter, we shall cover the following topics:

- Matrix creation
- Matrix operations
- Basic ufuncs
- Trigonometric functions
- Bitwise functions
- Comparison functions

Matrices

Matrices in NumPy are subclasses of `ndarray`. Matrices can be created using a special string format. They are, just like in mathematics, two-dimensional. Matrix multiplication is, as you would expect, different from the normal NumPy multiplication. The same is true for the power operator. We can create matrices with the `mat`, `matrix`, and `bmat` functions.

Time for action – creating matrices

Matrices can be created with the `mat` function. This function does not make a copy if the input is already a `matrix` or an `ndarray`. Calling this function is equivalent to calling `matrix(data, copy=False)`. We will also demonstrate transposing and inverting matrices.

1. **Creating a matrix from a string**: Rows are delimited by a semicolon, values by a space. Call the `mat` function with the following string to create a matrix:

    ```
    A = numpy.mat('1 2 3; 4 5 6; 7 8 9')
    print "Creation from string", A
    ```

    The matrix output should be the following matrix:

    ```
    Creation from string [[1 2 3]
     [4 5 6]
     [7 8 9]]
    ```

2. **Transposing matrices**: Transpose the matrix with the `T` attribute, as follows:

    ```
    print "transpose A", A.T
    ```

    The following is the transposed matrix:

    ```
    transpose A [[1 4 7]
     [2 5 8]
     [3 6 9]]
    ```

3. **Inverting matrices**: The matrix can be inverted with the `I` attribute, as follows:

    ```
    print "Inverse A", A.I
    ```

    The inverse matrix is as follows (be warned that this is a $O(n^3)$ operation):

    ```
    Inverse A [[ -4.50359963e+15   9.00719925e+15  -4.50359963e+15]
     [  9.00719925e+15  -1.80143985e+16   9.00719925e+15]
     [ -4.50359963e+15   9.00719925e+15  -4.50359963e+15]]
    ```

4. **Creating matrices from arrays**: Instead of using a string to create a matrix, let's do it with an array:

    ```
    print "Creation from array", numpy.mat(numpy.arange(9).reshape(3, 3))
    ```

    The newly-created array is as follows:

    ```
    Creation from array [[0 1 2]
     [3 4 5]
     [6 7 8]]
    ```

What just happened?

We created matrices with the `mat` function. We transposed the matrices with the `T` attribute and inverted them with the `I` attribute.

Creating a matrix from other matrices

Sometimes we want to create a matrix from other smaller matrices. We can do this with the `bmat` function. The `b` here stands for block matrix.

Time for action – creating a matrix from other matrices

We will create a matrix from two smaller matrices, as follows:

1. **Creating the smaller matrices**: First create a 2-by-2 identity matrix:

    ```
    A = numpy.eye(2)
    print "A", A
    ```

    The identity matrix looks like this:

    ```
    A [[ 1.  0.]
     [ 0.  1.]]
    ```

    Create another matrix like `A` and multiply by `2`:

    ```
    B = 2 * A
    print "B", B
    ```

    The second matrix is as follows:

    ```
    B [[ 2.  0.]
     [ 0.  2.]]
    ```

2. **Creating the compound matrix**: Create the compound matrix from a string. The string uses the same format as the `mat` function; only, you can use matrices instead of numbers.

    ```
    print "Compound matrix\n", numpy.bmat("A B; A B")
    ```

    The compound matrix is shown as follows:

    ```
    Compound matrix
    [[ 1.  0.  2.  0.]
     [ 0.  1.  0.  2.]
     [ 1.  0.  2.  0.]
     [ 0.  1.  0.  2.]]
    ```

What just happened?

We created a block matrix from two smaller matrices, with the `bmat` function. We gave the function a string containing the names of matrices instead of numbers.

Pop quiz – defining a matrix with a string

1. What is the row delimiter in a string accepted by the `mat` and `bmat` functions?

    a. Semicolon

    b. Colon

    c. Comma

    d. Space

Universal functions

Ufuncs expect a set of scalars as input and produce a set of scalars as output. Universal functions can typically be mapped to mathematical counterparts, such as, add, subtract, divide, multiply, and so on.

Time for action – creating universal function

We can create a universal function from a Python function with the NumPy `frompyfunc` function, as follows:

1. **Defining the Python function**: Define a Python function that answers the ultimate question to the universe, existence, and the rest (it's from a book, if you don't know which one, you can safely ignore this).

```
def ultimate_answer(a):
```

So far, nothing special; we gave the function the name `ultimate_answer` and defined one parameter, `a`.

2. **Initializing the result**: Create a result consisting of all zeros, that has the same shape as `a`, with the `zeros_like` function:

```
result = numpy.zeros_like(a)
```

3. **Complete the function**: Now set the elements of the initialized array to the answer `42` and return the result. The complete function should appear as shown, in the following code snippet. The `flat` attribute gives us access to a flat iterator that allows us to set the value of the array:

```
def ultimate_answer(a):
    result = numpy.zeros_like(a)
    result.flat = 42
    return result
```

4. **Create the universal function**: Create a universal function with `frompyfunc`; specify `1` as input and `1` as output:

```
ufunc = numpy.frompyfunc(ultimate_answer, 1, 1)
print "The answer", ufunc(numpy.arange(4))
```

The result for a one-dimensional array is shown as follows:

The answer [42 42 42 42]

We can do the same for a two-dimensional array by using the following code:

```
print "The answer", ufunc(numpy.arange(4).reshape(2, 2))
```

The output for a two dimensional array is shown as follows

The answer [[42 42]
[[42 42]
[42 42]]

What just happened?

We defined a Python function. In this function, we initialized to zero the elements of an array, based on the shape of an input argument, with the `zeros_like` function. Then, with the `flat` attribute of `ndarray`, we set the array elements to the ultimate answer, `42`.

Universal function methods

How can functions have methods? As we said earlier, universal functions are not functions but objects representing functions. Universal functions have four methods. They only make sense for functions such as `add`. That is, they have two input parameters and return one output parameter. If the signature of a ufunc does not match this condition, this will result in a `ValueError`, so call this method only for binary universal functions. The four methods are listed as follows:

1. `reduce`
2. `accumulate`
3. `reduceat`
4. `outer`

Time for action – applying the ufunc methods on add

Let's call the four methods on add.

1. **Calling the reduce method**: The input array is reduced by applying the universal function recursively along a specified axis on consecutive elements. For the add function, the result of reducing is similar to calculating the sum of an array. Call the reduce method:

```
a = numpy.arange(9)
print "Reduce", numpy.add.reduce(a)
```

The reduced array should be as follows:

```
Reduce 36
```

2. **Calling the accumulate method**: The accumulate method also recursively goes through the input array. But, contrary to the reduce method, it stores the intermediate results in an array and returns that. The result, in the case of the add function, is equivalent to calling the cumsum function. Call the accumulate method on the add function:

```
print "Accumulate", numpy.add.accumulate(a)
```

The accumulated array:

```
Accumulate [ 0  1  3  6 10 15 21 28 36]
```

3. **Calling the reduceat method**: The reduceat method is a bit complicated to explain, so let's call it and go through its algorithm, step-by-step. The reduceat method requires as arguments an input array and a list of indices:

```
print "Reduceat", numpy.add.reduceat(a, [0, 5, 2, 7])
```

The result is shown as follows:

```
Reduceat [10  5 20 15]
```

The first step concerns the indices 0 and 5. This step results in a reduce operation of the array elements between indices 0 and 5.

```
print "Reduceat step I", numpy.add.reduce(a[0:5])
```

The output of step 1 is as follows:

```
Reduceat step I 10
```

The second step concerns indices 5 and 2. Since 2 is less than 5, the array element at index 5 is returned:

```
print "Reduceat step II", a[5]
```

The second step results in the following output:

`Reduceat step II 5`

The third step concerns indices 2 and 7. This step results in a reduce operation of the array elements between indices 2 and 7:

`print "Reduceat step III", numpy.add.reduce(a[2:7])`

The result of the third step is shown as follows:

`Reduceat step III 20`

The fourth step concerns index 7. This step results in a reduce operation of the array elements from index 7 to the end of the array:

`print "Reduceat step IV", numpy.add.reduce(a[7:])`

The fourth step result is shown as follows:

`Reduceat step IV 15`

4. **Calling the outer method**: The `outer` method returns an array that has a rank which is the sum of the ranks of its two input arrays. The method is applied to all possible pairs of the input array elements. Call the `outer` method on the `add` function:

    `print "Outer", numpy.add.outer(numpy.arange(3), a)`

    The outer sum output result is as follows:

    ```
    Outer [[ 0  1  2  3  4  5  6  7  8]
     [ 1  2  3  4  5  6  7  8  9]
     [ 2  3  4  5  6  7  8  9 10]]
    ```

What just happened?

We applied the four methods, `reduce`, `accumulate`, `reduceat`, and `outer`, of universal functions to the `add` function. Since this is a binary function, no exception was thrown.

Arithmetic functions

The common arithmetic operators +, -, and * are implicitly linked to the `add`, `subtract`, and `multiply` universal functions. This means that when you use one of those operators on a NumPy array, the corresponding universal function will get called. Division involves a slightly more complex process. There are three universal functions that have to do with array division: `divide`, `true_divide`, and `floor_division`. Two operators correspond to division: / and //.

Time for action – dividing arrays

Let's see the array division in action:

1. **Calling divide**: The `divide` function does truncated integer division and normal floating-point division:

    ```
    a = numpy.array([2, 6, 5])
    b = numpy.array([1, 2, 3])
    print "Divide", numpy.divide(a, b), numpy.divide(b, a)
    ```

    The result of the `divide` function is shown as follows:

    ```
    Divide [2 3 1] [0 0 0]
    ```

    As you can see, big-time truncation takes place.

2. **Calling true_divide**: The `true_divide` function comes closer to the mathematical definition of division. Integer division returns a floating-point result and no truncation occurs:

    ```
    print "True Divide", numpy.true_divide(a, b), numpy.true_divide(b,
    a)
    ```

    The result of the `true_divide` function is as follows:

    ```
    True Divide [ 2.          3.          1.66666667] [ 0.5
    0.33333333   0.6        ]
    ```

3. **Calling floor_divide**: The `floor_divide` function always returns an integer result. It is equivalent to calling the `floor` function after calling the `divide` function. The `floor` function discards the decimal part of a floating-point number and returns an integer:

    ```
    print "Floor Divide", numpy.floor_divide(a, b), numpy.floor_
    divide(b, a)
    c = 3.14 * b
    print "Floor Divide 2", numpy.floor_divide(c, b), numpy.floor_
    divide(b, c)
    ```

    The `floor_divide` function results in:

    ```
    Floor Divide [2 3 1] [0 0 0]
    Floor Divide 2 [ 3.  3.  3.] [ 0.  0.  0.]
    ```

4. **Using the / operator**: By default, the / operator is equivalent to calling the `divide` function:

    ```
    from __future__ import division
    ```

    However, if this line is found at the beginning of a Python program, the `true_divide` function is called instead. So, this code would appear as follows:

    ```
    print "/ operator", a/b, b/a
    ```

The result is shown as follows:

```
/ operator [ 2.           3.          1.66666667] [ 0.5
0.33333333  0.6        ]
```

5. **Using the // operator**: The // operator is equivalent to calling the `floor_divide` function. For example, look at the following code snippet:

```
print "// operator", a//b, b//a
print "// operator 2", c//b, b//c
```

The // operator result is shown as follows:

```
// operator [2 3 1] [0 0 0]
// operator 2 [ 3.  3.  3.] [ 0.  0.  0.]
```

What just happened?

We found that there are three different NumPy division functions. The `divide` function truncates the integer division and normal floating-point division. The `true_divide` function always returns a floating-point result without any truncation. The `floor_divide` function always returns an integer result; the result is the same that you would get by calling the `divide` and `floor` functions consecutively.

Have a go hero – experimenting with __future__.division

Experiment to confirm the impact of importing __future__.division.

Modulo operation

The modulo or remainder can be calculated using the NumPy `mod`, `remainder`, and `fmod` functions. Also, one can use the `%` operator. The main difference among these functions is how they deal with negative numbers. The odd one out in this group is the `fmod` function.

Time for action – computing the modulo

Let's call the aforementioned functions:

1. **Calling the remainder function**: The `remainder` function returns the remainder of the two arrays, element-wise. 0 is returned if the second number is 0:

```
a = numpy.arange(-4, 4)
print "Remainder", numpy.remainder(a, 2)
```

The result of the `remainder` function is shown as follows:

```
Remainder [0 1 0 1 0 1 0 1]
```

2. **Calling the mod function**: The mod function does exactly the same as the remainder function:

```
print "Mod", numpy.mod(a, 2)
```

The result of the mod function is shown as follows:

```
Mod [0 1 0 1 0 1 0 1]
```

3. **Using the % operator**: The % operator is just shorthand for the remainder function:

```
print "% operator", a % 2
```

The result of the % operator is shown as follows:

```
% operator [0 1 0 1 0 1 0 1]
```

4. **Calling the fmod function**: The fmod function handles negative numbers differently than mod, fmod, and % do. The sign of the remainder is the sign of the dividend, and the sign of the divisor has no influence on the results:

```
print "Fmod", numpy.fmod(a, 2)
```

The fmod result is shown as follows:

```
Fmod [ 0 -1  0 -1  0  1  0  1]
```

What just happened?

We demonstrated the NumPy mod, remainder, and fmod functions, which compute the modulo, or remainder.

Fibonacci numbers

The Fibonacci numbers are based on a recurrence relation. It is difficult to express this relation directly with NumPy code. However, we can express this relation in a matrix form or use the golden ratio formula. This will introduce the matrix and rint functions. The matrix function creates matrices and the rint function rounds numbers to the closest integer, but the result is not integer.

Time for action – computing Fibonacci numbers

The Fibonacci recurrence relation can be represented by a matrix. Calculation of Fibonacci numbers can be expressed as repeated matrix multiplication:

1. **Creating the Fibonacci matrix**: Create the Fibonacci matrix as follows:

```
F = numpy.matrix([[1, 1], [1, 0]])
print "F", F
```

The Fibonacci matrix appears as follows:

```
F [[1 1]
 [1 0]]
```

2. **Computing a Fibonacci number with the matrix**: Calculate the 8th Fibonacci number (ignoring 0), by subtracting 1 from 8 and taking the power of the matrix. The Fibonacci number then appears on the diagonal:

```
print "8th Fibonacci", (F ** 7)[0, 0]
```

The Fibonacci number is:

```
8th Fibonacci 21
```

3. **Calculating with the golden ratio formula**: The golden ratio formula, better known as Binet's formula, allows us to calculate Fibonacci numbers with a rounding step at the end. Calculate the first eight Fibonacci numbers:

```
n = numpy.arange(1, 9)
sqrt5 = numpy.sqrt(5)
phi = (1 - sqrt5)/2
fibonacci = numpy.rint((phi**n - (-1/phi)**n)/sqrt5)
print "Fibonacci", fibonacci
```

The Fibonacci numbers are:

```
Fibonacci [  1.   1.   2.   3.   5.   8.  13.  21.]
```

What just happened?

We computed Fibonacci numbers in two ways. In the process, we learned about the `matrix` function that creates matrices. We also learned about the `rint` function that rounds numbers to the closest integer but does not change the type to integer.

Have a go hero – timing the calculations

You are probably wondering which approach is faster, so go ahead time it. Create a universal Fibonacci function with `frompyfunc` and time it too.

Lissajous curves

All the standard trigonometric functions, such as, sin, cos, tan, and so on are represented by universal functions in NumPy. Lissajous curves are a fun way of using trigonometry. I remember producing Lissajous figures on an oscilloscope in the physics lab. Two parametric equations can describe the figures:

$$x = A \sin(at + \pi/2)$$
$$y = B \sin(bt)$$

Time for action – drawing Lissajous curves

The Lissajous figures are determined by four parameters A, B, a, and b. Let's set A and B to 1 for simplicity:

1. **Initialize t**: Initialize t with the `linspace` function from -pi to pi with 201 points:

    ```
    a = float(sys.argv[1])
    b = float(sys.argv[2])
    t = numpy.linspace(-numpy.pi, numpy.pi, 201)
    ```

2. **Calculate x**: Calculate x with the `sin` function and `numpy.pi`:

    ```
    x = numpy.sin(a * t + numpy.pi/2)
    ```

3. **Calculate y**: Calculate y with the `sin` function:

    ```
    y = numpy.sin(b * t)
    ```

4. **Plot with Matplotlib**: `Matplotlib` will be covered later in *Chapter 9, Plotting with Matplotlib*. Plot as shown here:

    ```
    plot(x, y)
    show()
    ```

    The result for a = 9 and b = 8 :

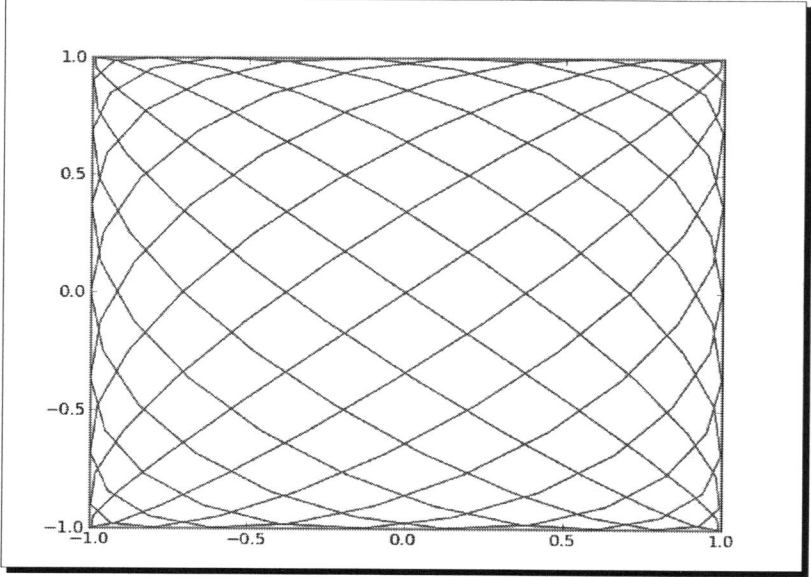

What just happened?

We plotted the Lissajous curve with the aforementioned parametric equations where A=B=1, a=9 and, b=8. We used the `sin` and `linspace` functions as well as the NumPy `pi` constant.

Square waves

Square waves are also one of those neat things that you can view on an oscilloscope. They can be approximated pretty well with sine waves; after all, a square wave is a signal that can be represented by an infinite Fourier series. The formula of the series is as follows:

$$\sum_{k=1}^{\infty} \frac{4\sin((2k-1)t)}{(2k-1)\pi}$$

Time for action – drawing a square wave

We will initialize t just like in the previous tutorial. We need to sum a number of terms. The higher the number of terms, the more accurate the result; k = 99 should be sufficient. In order to draw a square wave, follow the ensuing steps:

1. **Initialize t and k:** We will start by initializing t and k. Set initial values for the function to 0:

```
t = numpy.linspace(-numpy.pi, numpy.pi, 201)
k = numpy.arange(1, float(sys.argv[1]))
k = 2 * k - 1
f = numpy.zeros_like(t)
```

2. **Compute the function values:** This step should be a straightforward application of the `sin` and `sum` functions:

```
for i in range(len(t)):
    f[i] = numpy.sum(numpy.sin(k * t[i])/k)
f = (4 / numpy.pi) * f
```

3. **Plotting with Matplotlib:** The code to plot is almost identical to the one in the previous tutorial:

```
plot(t, f)
show()
```

The resulting square wave generated with k = 99 is as follows:

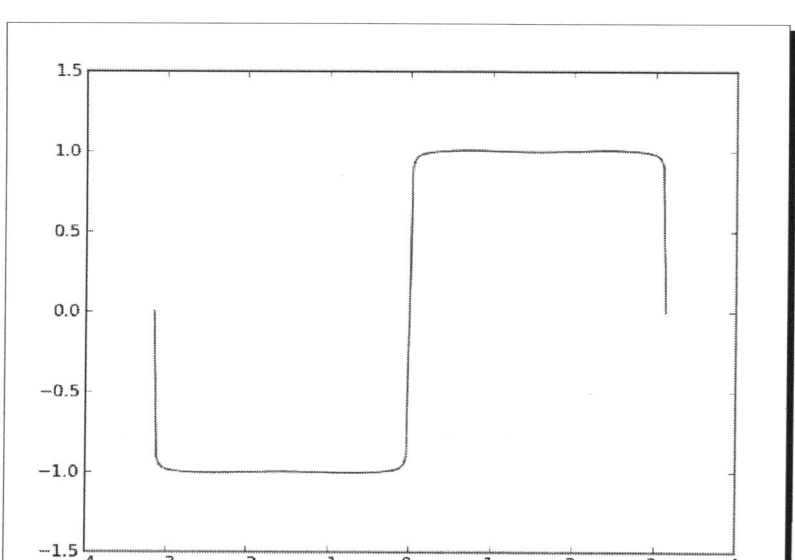

What just happened?

We generated a square wave or, at least, a fair approximation of it, using the `sin` function. The input values were assembled with `linspace` and the k values with the `arange` function.

Have a go hero – getting rid of the loop

You may have noticed that there is one loop in the code. Get rid of it with NumPy functions and make sure the performance is also improved.

Sawtooth and triangle waves

Sawtooth and triangle waves are also a phenomenon easily viewed on an oscilloscope. Just like with square waves, we can define an infinite Fourier series. The triangle waves can be found by taking the absolute value of a sawtooth wave. The formula for the representation of a series of sawtooth waves is:

$$\sum_{k=1}^{\infty} \frac{-2\sin(2\pi kt)}{k\pi}$$

Time for action – drawing sawtooth and triangle waves

We will initialize t just like in the previous tutorial. Again, k = 99 should be sufficient. In order to draw sawtooth and triangle waves, follow the ensuing steps:

1. **Initialize t and k**: Set initial values for the function to zero:

```
t = numpy.linspace(-numpy.pi, numpy.pi, 201)
k = numpy.arange(1, float(sys.argv[1]))
f = numpy.zeros_like(t)
```

Compute the function values: This should again be a straightforward application of the sin and sum functions:

```
for i in range(len(t)):
    f[i] = numpy.sum(numpy.sin(2 * numpy.pi * k * t[i])/k)
f = (-2 / numpy.pi) * f
```

2. **Plotting with Matplotlib**: It's easy to plot the sawtooth and triangle waves, since the value of the triangle wave should be equal to the absolute value of the sawtooth wave. Plot the waves as shown next:

```
plot(t, f, lw=1.0)
plot(t, numpy.abs(f), lw=2.0)
show()
```

In the following figure, the triangle wave is the one with the thicker line:

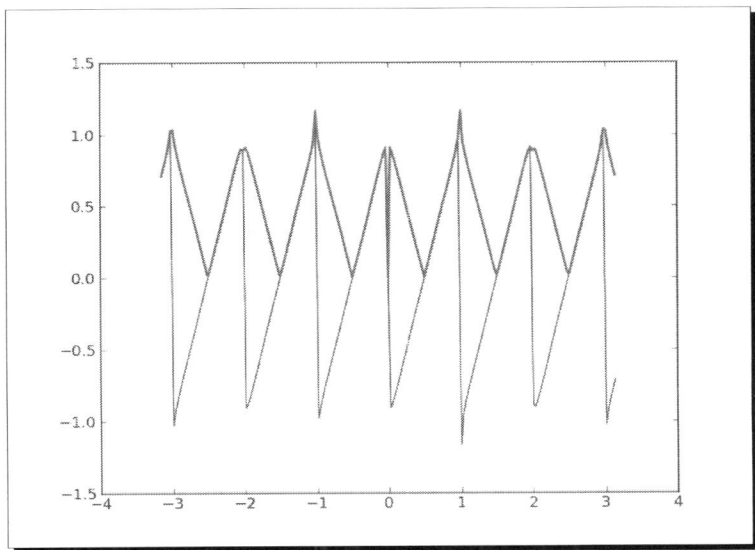

What just happened?

We drew a sawtooth wave using the `sin` function. The input values were assembled with `linspace` and the `k` values with the `arange` function. A triangle wave was derived from the sawtooth wave by taking the absolute value.

Have a go hero – getting rid of the loop

Your challenge, should you choose to accept it, is to get rid of the loop in the program. It should be doable with NumPy functions and the performance should double.

Bitwise and comparison functions

Bitwise functions operate on the bits of integers or integer arrays, since they are universal functions. The operators ^, &, |, <<, >>, and so on, have their NumPy counterparts The same goes for comparison operators, such as, <, >, ==, and so on. These operators allow you some clever tricks, which should be good for performance; however, they could make your code quite unreadable, so use them with care.

Time for action – twiddling bits

We will go over three tricks—checking whether the signs of integers are different, checking whether a number is a power of 2, and calculating the modulus of a number that is a power of 2. We will show an operators only notation and one using the corresponding NumPy functions:

1. **Checking signs:** The first trick depends on the XOR or ^ operator. The XOR operator is also called the inequality operator; so, if the sign bit of the two operands is different, the XOR operation will lead to a negative number. ^ corresponds to the `bitwise_xor` function. < corresponds to the `less` function.

    ```
    x = numpy.arange(-9, 9)
    y = -x
    print "Sign different?", (x ^ y) < 0
    print "Sign different?", numpy.less(numpy.bitwise_xor(x, y), 0)
    ```

    The result is shown as follows:

    ```
    Sign different? [ True  True  True  True  True  True  True  True
    True False  True  True
      True  True  True  True  True  True]
    Sign different? [ True  True  True  True  True  True  True  True
    True False  True  True
      True  True  True  True  True  True]
    ```

    As expected, all the signs differ, except for zero.

2. **'Power of 2' check**: A power of 2 is represented by a 1, followed by a series of trailing zeroes in binary notation. For instance, 10, 100, or 1000. A number one less than a power of 2 would be represented by a row of ones in binary. For instance, 11, 111, or 1111 (or 3, 7, and 15, in the decimal system). Now, if we bitwise AND a power of 2, and the integer that is one less than that, then we should get 0. The NumPy counterpart of & is `bitwise_and`; the counterpart of == is the `equal` universal function.

```
print "Power of 2?\n", x, "\n", (x & (x - 1)) == 0
print "Power of 2?\n", x, "\n", numpy.equal(numpy.bitwise_and(x,
    (x - 1)), 0)
```

The result is shown as follows:

```
Power of 2?
[-9 -8 -7 -6 -5 -4 -3 -2 -1  0  1  2  3  4  5  6  7  8]
[False False False False False False False False False  True  True
True
 False  True False False False  True]
Power of 2?
[-9 -8 -7 -6 -5 -4 -3 -2 -1  0  1  2  3  4  5  6  7  8]
[False False False False False False False False False  True  True
True
 False  True False False False  True]
```

3. **Computing the modulus of 4**: This trick actually works when taking the modulus of integers that are a power of 2, such as, 4, 8, 16, and so on. A bitwise left shift leads to doubling of values. We saw in the previous step that subtracting one from a power of 2 leads to a number in binary notation that has a row of ones, such as, 11, 111, or 1111. This basically gives us a mask. Bitwise-ANDing with such a number gives you the remainder with a power of 2. The NumPy equivalent of << is the `left_shift` universal function.

```
print "Modulus 4\n", x, "\n", x & ((1 << 2) - 1)
print "Modulus 4\n", x, "\n", numpy.bitwise_and(x,
    numpy.left_shift(1, 2) - 1)
```

The result is shown as follows:

```
Modulus 4
[-9 -8 -7 -6 -5 -4 -3 -2 -1  0  1  2  3  4  5  6  7  8]
[3 0 1 2 3 0 1 2 3 0 1 2 3 0 1 2 3 0]
Modulus 4
[-9 -8 -7 -6 -5 -4 -3 -2 -1  0  1  2  3  4  5  6  7  8]
[3 0 1 2 3 0 1 2 3 0 1 2 3 0 1 2 3 0]
```

What just happened?

We covered three bit-twiddling hacks—checking whether the signs of integers are different, checking whether a number is a power of 2, and calculating the modulus of a number that is a power of 2. We saw the NumPy counterparts of the operators ^, &, <<, and <.

Summary

We learned, in this chapter, about matrices and universal functions. We covered how to create matrices and how universal functions work. We had a brief introduction to arithmetic, trigonometric, bitwise, and comparison universal functions.

In the next chapter, we shall cover NumPy modules.

Move Further with NumPy Modules

NumPy has a number of modules that have been inherited from its predecessor, Numeric . Some of these packages have a SciPy counterpart, which may have fuller functionality. This will be discussed in a later chapter. The `numpy.dual` *package contains functions that are defined both in NumPy and SciPy. The packages discussed in this chapter are also part of the* `numpy.dual` *package.*

In this chapter, we shall cover the following topics:

◆ The `linalg` package

◆ The `fft` package

◆ Random numbers

◆ Continuous and discrete distributions

Linear algebra

The `numpy.linalg` package contains linear algebra functions. With this module, you can invert matrices, calculate eigenvalues, solve linear equations, and determine determinants among other things.

Time for action – inverting matrices

The inverse of a matrix A in linear algebra is the matrix A^{-1}, which, when multiplied with the original matrix, is equal to the identity matrix I. This can be written as follows:

$A \ A^{-1} = I$

The `inv` function in the `numpy.linalg` package can do this for us. Let's invert an example matrix. To invert matrices, follow the ensuing steps:

1. **Create the example matrix**: We will create the example matrix with the `mat` function that we used in previous chapters.

```
A = numpy.mat("0 1 2;1 0 3;4 -3 8")
print "A\n", A
```

The A matrix is shown as follows:

```
A
[[ 0   1   2]
 [ 1   0   3]
 [ 4  -3   8]]
```

2. **Invert the matrix**: Now, we can see the `inv` function in action.

```
inverse = numpy.linalg.inv(A)
print "inverse of A\n", inverse
```

The inverse matrix is shown as follows:

```
inverse of A
[[-4.5   7.    -1.5]
 [-2.    4.    -1. ]
 [ 1.5  -2.     0.5]]
```

If the matrix is singular, or not square, a `LinAlgError` is raised. If you want, you can check the result manually. This is left as an exercise for the reader.

3. **Check by multiplication**: Let's check what we get when we multiply the original matrix with the result of the `inv` function:

```
print "Check\n", A * inverse
```

The result is the identity matrix, as expected.

```
Check
[[ 1.   0.   0.]
 [ 0.   1.   0.]
 [ 0.   0.   1.]]
```

What just happened?

We calculated the inverse of a matrix with the `inv` function of the `numpy.linalg` package. We checked, with matrix multiplication, whether this is indeed the inverse matrix.

Pop quiz – creating a matrix

1. Which function can create matrices?

    a. array

    b. create_matrix

    c. mat

    d. vector

Have a go hero – inverting your own matrix

Create your own matrix and invert it. The inverse is only defined for square matrices. The matrix must be square and invertible; otherwise, a `LinAlgError` exception is raised.

Solving linear systems

The `numpy.linalg` function `solve` solves systems of linear equations of the form $Ax = b$; here A is a matrix, b can be 1D or 2D array, and x is an unknown variable. We will see the `dot` function in action. This function returns the dot product of two floating-point arrays.

Time for action – solving a linear system

Let's solve an example of linear system. To solve a linear system, follow the ensuing steps:

1. **Create the matrices A and b**: Let's create A and b:

   ```
   A = numpy.mat("1 -2 1;0 2 -8;-4 5 9")
   print "A\n", A
   b = numpy.array([0, 8, -9])
   print "b\n", b
   ```

   The matrices A and b are shown as follows:

   ```
   A
   [[ 1 -2  1]
    [ 0  2 -8]
    [-4  5  9]]
   b [ 0  8 -9]
   ```

2. **Call the solve function**: Solve this linear system with the `solve` function:

   ```
   x = numpy.linalg.solve(A, b)
   print "Solution", x
   ```

   The solution of the linear system is as follows:

   ```
   Solution [ 29.  16.   3.]
   ```

3. **Check with the dot function**: Check whether the solution is correct with the dot function:

```
print "Check\n", numpy.dot(A , x)
```

The result is as expected:

```
Check
[[ 0.   8.  -9.]]
```

What just happened?

We solved a linear system using the `solve` function from the NumPy `linalg` module and checked the solution with the `dot` function.

Finding eigenvalues and eigenvectors

Eigenvalues are scalar solutions to the equation Ax = ax, where A is a two-dimensional matrix and x is a one-dimensional vector. **Eigenvectors** are vectors corresponding to eigenvalues. The `eigvals` function in the `numpy.linalg` package calculates eigenvalues. The `eig` function returns a tuple containing eigenvalues and eigenvectors.

Time for action – determining eigenvalues and eigenvectors

Let's calculate the eigenvalues of a matrix:

1. **Create the matrix**: Create a matrix as shown.

```
A = numpy.mat("3 -2;1 0")
print "A\n", A
```

The matrix we created looks like this:

```
A
[[ 3 -2]
 [ 1  0]]
```

2. **Calculate eigenvalues with the eig function**: Call the `eig` function.

```
print "Eigenvalues", numpy.linalg.eigvals(A)
```

The eigenvalues of the matrix are as follows:

```
Eigenvalues [ 2.   1.]
```

3. **Getting eigenvalues and eigenvectors with eig**: Determine eigenvalues and eigenvectors with the `eig` function. This function returns a tuple, where the first element contains eigenvalues and the second element contains corresponding `eigenvectors`, arranged column-wise.

```
eigenvalues, eigenvectors = numpy.linalg.eig(A)
print "First tuple of eig", eigenvalues
print "Second tuple of eig\n", eigenvectors
```

The eigenvalues and eigenvectors will be:

```
First tuple of eig [ 2.   1.]
Second tuple of eig
[[ 0.89442719  0.70710678]
 [ 0.4472136   0.70710678]]
```

4. **Check the result**: Check the result with the dot function by calculating the right and left side of the eigenvalues equation Ax = ax.

```
for i in range(len(eigenvalues)):
    print "Left", numpy.dot(A, eigenvectors[:,i])
    print "Right", eigenvalues[i] * eigenvectors[:,i]
    print
```

The output is as follows:

```
Left [[ 1.78885438]
 [ 0.89442719]]
Right [[ 1.78885438]
 [ 0.89442719]]
Left [[ 0.70710678]
 [ 0.70710678]]
Right [[ 0.70710678]
 [ 0.70710678]]
```

What just happened?

We found the eigenvalues and eigenvectors of a matrix with the eigvals and eig functions of the numpy.linalg module. We checked the result using the dot function.

Singular value decomposition

Singular value decomposition is a type of factorization that decomposes a matrix into a product of three matrices. The svd function in the numpy.linalg package can perform this decomposition. This function returns two orthogonal matrices and the singular values of the middle matrix:

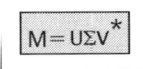

$$M = U\Sigma V^*$$

Time for action – decomposing a matrix

It's time to decompose a matrix with the singular value decomposition. In order to decompose a matrix, follow the ensuing steps:

1. **Create a matrix**: First, create a matrix as shown.

```
A = numpy.mat("4 11 14;8 7 -2")
print "A\n", A
```

The matrix we created looks like this:

```
A
[[ 4 11 14]
 [ 8  7 -2]]
```

2. **Decompose the matrix**: Decompose the matrix with the svd function.

```
U, Sigma, V = numpy.linalg.svd(A, full_matrices=False)
print "U"
print U
print "Sigma"
print Sigma
print "V"
print V
```

The result is a tuple containing the two orthogonal matrices U and V on the left and right and the singular values of the middle matrix.

```
U
[[-0.9486833  -0.31622777]
 [-0.31622777  0.9486833 ]]
Sigma
[ 18.97366596   9.48683298]
V
[[-0.33333333 -0.66666667 -0.66666667]
 [ 0.66666667  0.33333333 -0.66666667]]
```

3. **Check the decomposition by matrix multiplication**: We do not actually have the middle matrix—we only have the diagonal values. The other values are all 0. We can form the middle matrix with the diag function. Multiply the three matrices. This is shown as follows:

```
print "Product\n", U * numpy.diag(Sigma) * V
```

The product of the three matrices looks like this:

```
Product
[[  4.  11.  14.]
 [  8.   7.  -2.]]
```

What just happened?

We decomposed a matrix and checked the result by matrix multiplication. We used the `svd` function from the NumPy `linalg` module.

Pseudo inverse

The Moore-Penrose pseudo inverse of a matrix can be computed with the `pinv` function of the `numpy.linalg` module (see http://en.wikipedia.org/wiki/Moore%E2%80%93Penrose_pseudoinverse). The pseudo inverse is calculated using its singular value decomposition. The `inv` function only accepts square matrices; the `pinv` function does not have this restriction.

Time for action – computing the pseudo inverse of a matrix

Let's compute the pseudo inverse of a matrix:

1. **Create a matrix**: First, create a matrix as shown.

   ```
   A = numpy.mat("4 11 14;8 7 -2")
   print "A\n", A
   ```

   The matrix we created looks like this:

   ```
   A
   [[ 4 11 14]
    [ 8  7 -2]]
   ```

2. **Compute the pseudo inverse**: Calculate the pseudo inverse matrix with the `pinv` function as shown.

   ```
   pseudoinv = numpy.linalg.pinv(A)
   print "Pseudo inverse\n", pseudoinv
   ```

   The pseudo inverse is as follows:

   ```
   Pseudo inverse
   [[-0.00555556  0.07222222]
    [ 0.02222222  0.04444444]
    [ 0.05555556 -0.05555556]]
   ```

3. **Multiply the matrices**: Multiply the original and pseudo inverse matrices.

   ```
   print "Check", A * pseudoinv
   ```

   What we get is not an identity matrix, but it comes close to it:

   ```
   Check [[  1.00000000e+00   0.00000000e+00]
    [  8.32667268e-17   1.00000000e+00]]
   ```

What just happened?

We computed the pseudo inverse of a matrix with the `pinv` function of the `numpy.linalg` module. The check by matrix multiplication resulted in a matrix that is approximately an identity matrix.

Determinants

The **determinant** is a value associated with a matrix. It is used throughout mathematics. The `numpy.linalg` module has a `det` function that returns the determinant of a matrix.

Time for action – calculating the determinant of a matrix

To calculate the determinant of a matrix, follow the ensuing steps:

1. **Create a matrix**: Create the matrix as shown:

```
A = numpy.mat("3 4;5 6")
print "A\n", A
```

The matrix we created is shown as follows:

```
A
[[ 3.   4.]
 [ 5.   6.]]
```

2. **Determine the determinant**: Compute the determinant with the `det` function:

```
print "Determinant", numpy.linalg.det(A)
```

The determinant is shown as follows:

```
Determinant -2.0
```

What just happened?

We calculated the determinant of a matrix with the `det` function from the `numpy.linalg` module.

Fast Fourier transform

NumPy has a module called `fft` that offers fast Fourier transform functionality. A lot of the functions in this module are paired; this means that, for many functions, there is a function that does the inverse operation. For instance, the `fft` and `ifft` function.

Time for action – calculating the Fourier transform

First, we will create a signal to transform. In order to calculate the Fourier transform, follow the ensuing steps:

1. **Create the input signal**: Create a cosine wave with 30 points, as follows.

```
x =   numpy.linspace(0, 2 * numpy.pi, 30)
wave = numpy.cos(x)
```

2. **Transform the signal**: Transform the cosine wave with the `fft` function.

```
transformed = numpy.fft.fft(wave)
```

3. **Apply the inverse transform**: Apply the inverse transform with the `ifft` function. It should approximately return the original signal.

```
print numpy.all(numpy.abs(numpy.fft.ifft(transformed) - wave) < 10
** -9)
```

The result is shown as follows:

True

4. **Plot the transform**: Plot the transformed signal with Matplotlib:

```
plot(transformed)
show()
```

The resulting diagram shows the fast Fourier Transform:

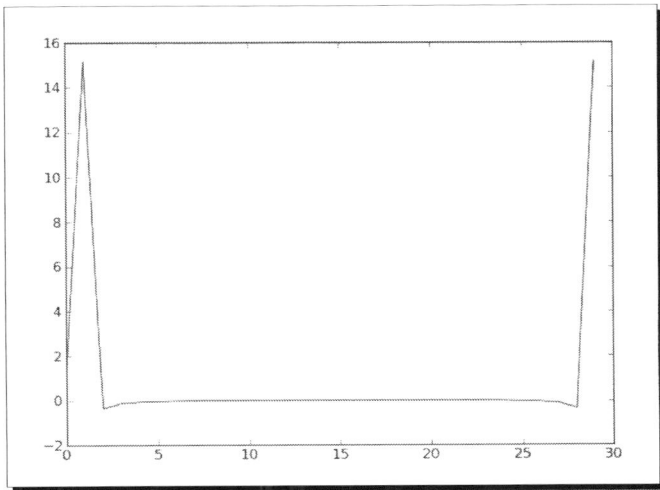

What just happened?

We applied the `fft` function to a cosine wave. After applying the `ifft` function, we got our signal back.

Shifting

The `fftshift` function of the `numpy.linalg` module shifts zero-frequency components to the center of a spectrum. The `ifftshift` function reverses this operation.

Time for action – shifting frequencies

We will create a signal, transform it, and then shift the signal. In order to shift the frequencies, follow the ensuing steps:

1. **Create the input signal**: Create a cosine wave with 30 points.

   ```
   x =  numpy.linspace(0, 2 * numpy.pi, 30)
   wave = numpy.cos(x)
   ```

2. **Transform the signal**: Transform the cosine wave with the `fft` function.

   ```
   transformed = numpy.fft.fft(wave)
   ```

3. **Shift the signal**: Shift the signal with the `fftshift` function.

   ```
   shifted = numpy.fft.fftshift(transformed)
   ```

4. **Reverse the shift**: Reverse the shift with the `ifftshift` function. This should undo the shift.

   ```
   print numpy.all((numpy.fft.ifftshift(shifted) - transformed) < 10
   ** -9)
   ```

   The result is shown as follows:

   True

5. **Plot**: Plot the signal and transform it with Matplotlib.

   ```
   plot(transformed, lw=2)
   plot(shifted, lw=3)
   show()
   ```

The following diagram shows the shift in the fast Fourier transform:

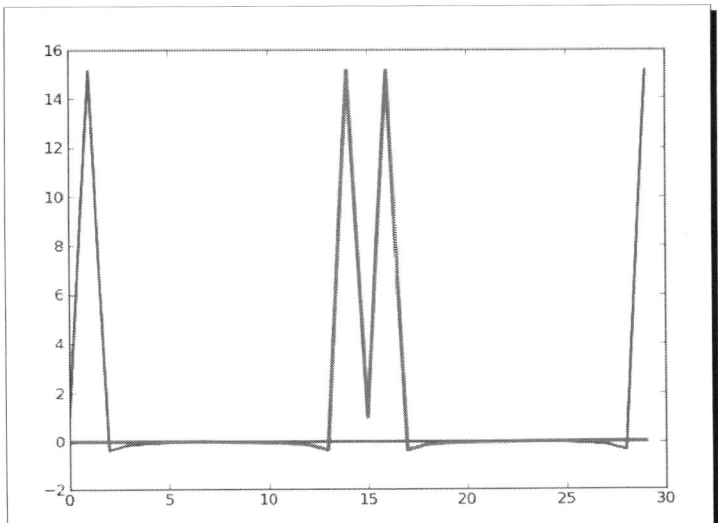

What just happened?

We applied the `fftshift` function to a cosine wave. After applying the `ifftshift` function, we got our signal back.

Random numbers

The random numbers related functions can be found in the NumPy `random` module. The core random number generator is based on the Mersenne Twister algorithm. Random numbers can be generated from discrete or continuous distributions. The distribution functions have an optional `size` parameter, which tells NumPy how many numbers to generate. You can specify either an integer or a tuple as size. This will result in an array filled with random numbers of appropriate shape. Discrete distributions include the geometric, hypergeometric, and binomial distributions.

Time for action – gambling with the binomial

The binomial distribution models the number of successes in an integer number of independent trials of an experiment, where the probability of success in each experiment is a fixed number. Imagine a 17th-century gambling house where you can bet on flipping of pieces of eight. Nine coins are flipped. If less than five are heads, then you lose one piece of eight, otherwise you win one. Let's simulate this, starting with 1000 coins in our possession. We will use the `binomial` function from the `random` module for that purpose.

In order to understand the `binomial` function, look at the following section:

1. **Calling the binomial function**: Initialize an array, which represents the cash balance, to zeros. Call the `binomial` function with a size of `10000`. This represents 10000 coin flips in our casino.

```
cash = numpy.zeros(10000)
cash[0] = 1000
outcome = numpy.random.binomial(9, 0.5, size=len(cash))
```

2. **Updating the cash balance**: Go through the outcomes of the coin flips and update the `cash` array. Print the minimum and maximum of the `outcome`, just to make sure we don't have any strange outliers.

```
for i in range(1, len(cash)):
   if outcome[i] < 5:
      cash[i] = cash[i - 1] - 1
   elif outcome[i] < 10:
      cash[i] = cash[i - 1] + 1
   else:
      raise AssertionError("Unexpected outcome " + outcome)
print outcome.min(), outcome.max()
```

As expected, the values are between 0 and 9:

```
0 9
```

3. **Plot**: Plot the cash array with Matplotlib:

```
plot(numpy.arange(len(cash)), cash)
show()
```

As you can see in the following diagram, our cash balance performs a random walk:

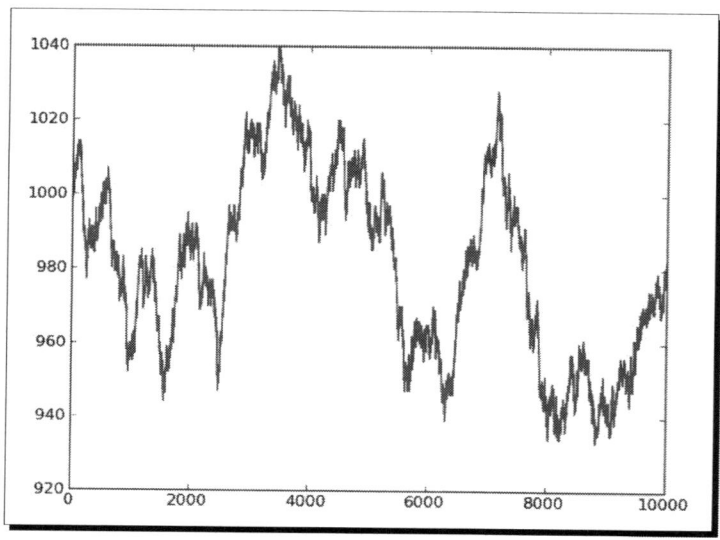

What just happened?

We did a random walk experiment using the `binomial` function from the NumPy random module.

Hypergeometric distribution

The **hypergeometric distribution** models a jar with two types of objects in it. The model tells us how many objects of one type we can get if we take a specified number of items out of the jar without replacing them. The NumPy `random` module has a `hypergeometric` function that simulates this situation.

Time for action – simulating a game show

Imagine a game show where every time the contestants answer a question correctly, they get to pull three balls from a jar and then put them back. Now there is a catch, there is one ball in there that is bad. Every time it is pulled out the contestants lose six points. If however, they manage to get out three of the twenty-five normal balls, they get one point. So, what is going to happen if we have a 100 questions in total? In order to get a solution for this, look at the following section:

1. **Initialize the outcomes of the game**: Initialize the outcome of the game with the `hypergeometric` function.

```
points = numpy.zeros(100)
outcomes = numpy.random.hypergeometric(25, 1, 3, size=len(points))
```

2. **Simulate the game**: Set the scores based on the `outcomes` from the previous step.

```
for i in range(len(points)):
   if outcomes[i] == 3:
      points[i] = points[i - 1] + 1
   elif outcomes[i] == 2:
      points[i] = points[i - 1] - 6
   else:
      print outcomes[i]
```

3. **Plot the points**: Plot the `points` array with Matplotlib.

```
plot(numpy.arange(len(points)), points)
show()
```

The following diagram shows how the scoring evolved:

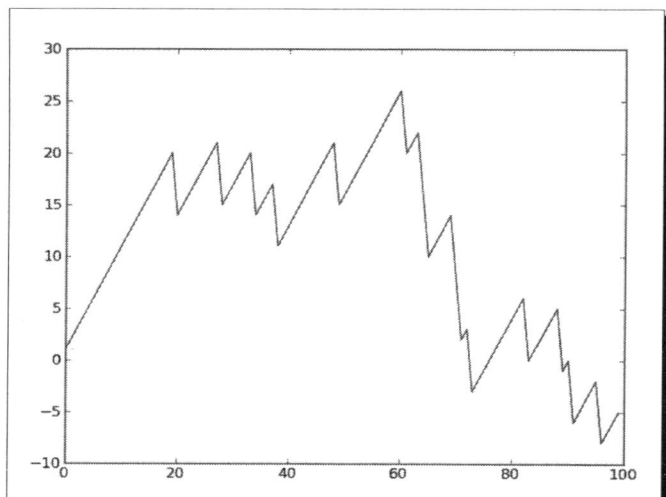

What just happened?

We simulated a game show using the `hypergeometric` function from the NumPy `random` module. The game scoring depends on how many good and how many bad balls are pulled out of a jar in each session.

Continuous distributions

Continuous distributions are modeled by the probability density functions (pdf). The probability for a certain interval is determined by integration of the probability density function. The NumPy `random` module has a number of functions that represent continuous distributions—`beta`, `chisquare`, `exponential`, `f`, `gamma`, `gumbel`, `laplace`, `lognormal`, `logistic`, `multivariate_normal`, `noncentral_chisquare`, `noncentral_f`, `normal`, and others.

Time for action – drawing a normal distribution

Random numbers can be generated from a normal distribution and their distribution may be visualized with a histogram. To draw a normal distribution, follow the ensuing steps:

1. **Generate values**: Generate random numbers using the `normal` function from the `random` NumPy module.

    ```
    N=10000
    normnal_values = numpy.random.normal(size=N)
    ```

2. **Draw the histogram and theoretical pdf**: Draw the histogram and theoretical pdf with a center value of 0 and standard deviation of 1. We will use Matplotlib for this purpose.

```
dummy, bins, dummy = matplotlib.pyplot.hist(normal_values,
   numpy.sqrt(N), normed=True, lw=1)
sigma = 1
mu = 0
matplotlib.pyplot.plot(bins, 1/(sigma * numpy.sqrt(2 * numpy.pi))
   * numpy.exp( - (bins - mu)**2 / (2 * sigma**2) ),lw=2)
matplotlib.pyplot.show()
```

In the following diagram, we see the familiar bell curve:

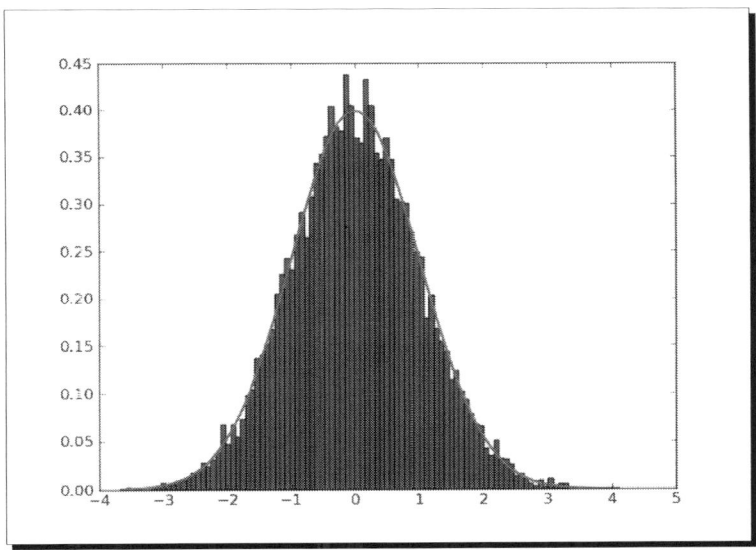

What just happened?

We visualized the normal distribution using the `normal` function from the `random` NumPy module. We did this by drawing the bell curve and a histogram of randomly-generated values.

Lognormal distribution

A **lognormal distribution** is a distribution of a variable whose natural logarithm is normally distributed. The `lognormal` function of the `random` NumPy module models this distribution.

Time for action – drawing the lognormal distribution

Let's visualize the lognormal distribution and its probability density function with a histogram:

1. **Generate**: Generate random numbers using the `normal` function from the `random` NumPy module.

```
N=10000
lognormal_values = numpy.random.lognormal(size=N)
```

2. **Draw the histogram and theoretical pdf**: Draw the histogram and theoretical pdf with a center value of 0 and standard deviation of 1. We will use Matplotlib for this purpose:

```
dummy, bins, dummy = matplotlib.pyplot.hist(lognormal_values,
    numpy.sqrt(N), normed=True, lw=1)
sigma = 1
mu = 0
x = numpy.linspace(min(bins), max(bins), len(bins))
pdf = numpy.exp(-(numpy.log(x) - mu)**2 / (2 * sigma**2))/ (x *
    sigma * numpy.sqrt(2 * numpy.pi))
matplotlib.pyplot.plot(x, pdf,lw=3)
matplotlib.pyplot.show()
```

The fit of the histogram and theoretical pdf is excellent, as you can see in the following diagram:

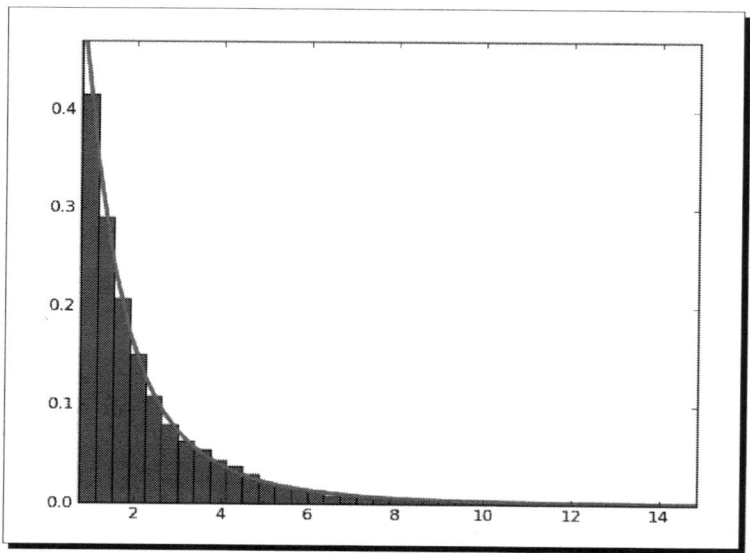

What just happened?

We visualized the lognormal distribution using the `lognormal` function from the `random` NumPy module. We did this by drawing the curve of the `theoretical probability density` function and a histogram of randomly-generated values.

Summary

We learned a lot in this chapter about NumPy modules. We covered linear algebra, the Fast Fourier transform, continuous and discrete distributions, and random numbers.

In the next chapter, we shall cover specialized routines. These are functions that you probably would not use often, but are very useful when you do need them.

7
Peeking Into Special Routines

As NumPy users, we sometimes find ourselves having special needs. Fortunately, NumPy provides for most of our needs. This chapter describes some of the more specialized NumPy functions.

In this chapter we will cover the following topics:

- Sorting and searching
- Special functions
- Financial utilities
- Window functions

Sorting

NumPy has several sorting routines:

- The `sort` function returns a sorted array
- The `lexsort` function performs sorting with a list of keys
- The `argsort` function returns the indices that would sort an array
- The `ndarray` class has a `sort` method that performs in place sorting
- The `msort` function sorts an array along the first axis
- The `sort_complex` function sorts complex numbers by their real part and then their imaginary part

Time for action – sorting lexically

The NumPy `lexsort` function returns an array of indices corresponding to lexically sorting an array. We need to give the function an array or tuple of sort keys:

1. **Loading the data**: Now for something completely different. Let's go back to *Chapter 3, Get into Terms with Commonly Used Functions.* In that chapter we used stock price data of AAPL. This is by now pretty old data. We will load the close prices and the always complex dates. In fact, we will need a converter function just for the dates:

```
def datestr2num(s):
    return datetime.datetime.strptime(s, "%d-%m-%Y").toordinal()

dates,closes=numpy.loadtxt('AAPL.csv', delimiter=',',
    usecols=(1, 6), converters={1:datestr2num}, unpack=True)
```

2. **Sorting lexically**: Sort the names lexically with the `lexsort` function. The data is already sorted by date, but we will now sort it by close as well:

```
indices = numpy.lexsort((dates, closes))

print "Indices", indices
print ["%s %s" % (datetime.date.fromordinal(dates[i]),
    closes[i]) for i in indices]
```

The code prints:

```
['2011-01-28 336.1', '2011-02-22 338.61', '2011-01-31 339.32',
'2011-02-23 342.62', '2011-02-24 342.88', '2011-02-03 343.44',
'2011-02-02 344.32', '2011-02-01 345.03', '2011-02-04 346.5',
'2011-03-10 346.67', '2011-02-25 348.16', '2011-03-01 349.31',
'2011-02-18 350.56', '2011-02-07 351.88', '2011-03-11 351.99',
'2011-03-02 352.12', '2011-03-09 352.47', '2011-02-28 353.21',
'2011-02-10 354.54', '2011-02-08 355.2', '2011-03-07 355.36',
'2011-03-08 355.76', '2011-02-11 356.85', '2011-02-09 358.16',
'2011-02-17 358.3', '2011-02-14 359.18', '2011-03-03 359.56',
'2011-02-15 359.9', '2011-03-04 360.0', '2011-02-16 363.13']
```

What just happened?

We sorted the close prices of AAPL lexically using the NumPy `lexsort` function. The function returned the indices corresponding with sorting the array.

Have a go hero – trying a different sort order

We sorted using the dates, close price sort order. Try a different order. Generate random numbers using the `random` module we learned about in the previous chapter and sort those using `lexsort`.

Complex numbers

Complex numbers are numbers that have a real and imaginary part. As you remember from previous chapters, NumPy has special complex datatypes that represent complex numbers by two floating point numbers. These numbers can be sorted using the NumPy `sort_complex` function. This function sorts the real part first and then the imaginary part.

Time for action – sorting complex numbers

We will create an array of complex numbers and sort it:

1. **Generating random complex numbers**: Generate five random numbers for the real part of the complex numbers and five numbers for the imaginary part. Seed the random generator to 42:

```
numpy.random.seed(42)
complex_numbers = numpy.random.random(5) + 1j * numpy.random.random(5)
print "Complex numbers\n", complex_numbers
```

2. **Calling sort_complex on the random numbers**: Call the `sort_complex` function to sort the complex numbers we generated in the previous step:

```
print "Sorted\n", numpy.sort_complex(complex_numbers)
```

The sorted numbers would be:

```
Sorted
[ 0.39342751+0.34955771j  0.40597665+0.77477433j
0.41516850+0.26221878j
  0.86631422+0.74612422j  0.92293095+0.81335691j]
```

What just happened?

We generated random complex numbers and sorted them using the `sort_complex` function.

Pop quiz – generating random numbers

Which NumPy module deals with random numbers?

◆ Randnum

◆ random

◆ randomutil

◆ rand

Searching

NumPy has several functions that can search through arrays:

◆ The `argmax` function gives the indices of the maximum values of an array.

◆ The `nanargmax` function does the same but ignores NaN values.

◆ The `argmin` and `nanargmin` functions provide similar functionality but pertaining to minimum values.

◆ The `argwhere` function searches for non-zero values and returns the corresponding indices grouped by element.

◆ The `searchsorted` function tells you the index in an array where a specified value could be inserted to maintain the sort order. It uses binary search, which is a O(log n) algorithm.

◆ The `extract` function retrieves values from an array based on a condition.

Time for action – using searchsorted

The `searchsorted` function allows us to get the index of a value in a sorted array, where it could be inserted so that the array remains sorted. An example should make this clear:

1. **Creating a sorted array**: To demonstrate we will need an array that is sorted. Create an array with `arange`, which of course is sorted.

```
a = numpy.arange(5)
```

2. **Calling searchsorted**: Time to call the `searchsorted` function.

```
indices = numpy.searchsorted(a, [-2, 7])
print "Indices", indices
```

The indices that should maintain the sort order.

```
Indices [0 5]
```

3. **Constructing the full array**: Let's construct the full array with the `insert` function.

```
print "The full array", numpy.insert(a, indices, [-2, 7])
```

This gives us the full array:

```
The full array [-2  0  1  2  3  4  7]
```

What just happened?

The `searchsorted` function gave us indices 5 and 0 for 7 and -2. With these indices, we would make the array `[-2, 0, 1, 2, 3, 4, 7]` — so the array remains sorted.

Array elements extraction

The NumPy `extract` function allows us to extract items from an array based on a condition.

Time for action – extracting elements from an array

Lets' extract the even elements from an array:

1. Create the array with the `arange` function:

```
a = numpy.arange(7)
```

2. Create the condition that selects the even elements:

```
condition = (a % 2) == 0
```

3. Extract the even elements based on our condition with the `extract` function:

```
print "Even numbers", numpy.extract(condition, a)
```

Giving us the even numbers as required:

```
Even numbers [0 2 4 6]
```

What just happened?

We extracted the even elements from an array based on a Boolean condition with the NumPy `extract` function.

Financial functions

NumPy has a number of financial utilities functions:

- The `fv` function calculates the so called future value
- The `pv` function computes the present value
- The `npv` function returns the net present value
- The `pmt` function computes the payment against loan principal plus interest
- The `irr` function calculates the internal rate of return
- The `mirr` function calculates the modified internal rate of return
- The `nper` function returns the number of periodic payments
- The `rate` function calculates the rate of interest

Time for action – determining future value

The future value depends on four parameters—the interest rate, the number of periods, a periodic payment, and the present value. In this tutorial, let's take an interest rate of 3 percent, quarterly payment of 10 for 5 years and present value of 1000:

1. **Calculating the future value**: Call the `fv` function with the appropriate values:

```
print "Future value", numpy.fv(0.03/4, 5 * 4, -10, -1000)
```

The future value is:

```
Future value 1376.09633204
```

What just happened?

We calculated the future value using the NumPy `fv` function starting with a present value of 1000, interest rate of 3 percent and quarterly payments of 10 for 5 years.

Present value

The NumPy `pv` function can calculate the present value. This function mirrors the `fv` function and requires the interest rate, number of periods, and the periodic payment as well, but here we start with the future value.

Time for action – getting the present value

Let's reverse— compute the present value with numbers from the previous tutorial:

1. **Calculating the present value**: Plug in the figures from the previous *Time for action* tutorial.

```
print "Present value", numpy.pv(0.03/4, 5 * 4, -10, 1376.09633204)
```

This gives us 1000 as expected apart from a tiny numerical error. Actually it is not an error but a representation issue. We are dealing here with outgoing cash flow, that is the reason for the negative value:

```
Present value -999.999999999
```

What just happened?

We did the reverse computation of the previous *Time for action* tutorial to get the present value from the future value. This was done with the NumPy `pv` function.

Net present value

The NumPy `npv` function returns the net present value of cash flows. The function requires two arguments, the rate and an array representing the cash flows.

Time for action – calculating the net present value

We will calculate the net present value for a random generated cash flow series:

1. **Generate the random cash flow series**: Generate five random values for the cash flow series. Insert -100 as start value.

```
cashflows = numpy.random.randint(100, size=5)
cashflows = numpy.insert(cashflows, 0, -100)
print "Cashflows", cashflows
```

The cash flows would be:

```
Cashflows [-100    38    48    90    17    36]
```

2. **Calculating net present value**: Call the `npv` function to calculate the net present value from the cash flow series we generated in the previous step. Use a rate of 3 percent.

```
print "Net present value", numpy.npv(0.03, cashflows)
```

The net present value:

```
Net present value 107.435682443
```

What just happened?

We computed the net present value from a random generated cash flow series with the NumPy `npv` function.

Internal rate of return

The NumPy `irr` function returns the internal rate of return for a given cash flow series.

Time for action – determining the internal rate of return

Let's reuse the cash flow series from the previous *Time for action* tutorial:

1. **Calling the irr function:** Call the `irr` function with the cash flow series from the previous *Time for action* tutorial:

```
print "Internal rate of return", numpy.irr([-100, 38, 48, 90,
    17, 36])
```

The internal rate of return:

```
Internal rate of return 0.373420226888
```

What just happened?

We calculated the internal rate of return from the cash flow series of the previous *Time for action* tutorial. The value was given by the NumPy `irr` function.

Periodic payments

The NumPy `pmt` function allows you to compute periodic payments for a loan based on an interest rate and the number of periodic payments.

Time for action – calculating the periodic payments

Suppose you have a loan of 1 million with interest rate of 10 percent. You have 30 years to pay the loan back. How much do you have to pay each month? Let's find out:

1. Call the `pmt` function with the values mentioned above:

```
print "Payment", numpy.pmt(0.01/12, 12 * 30, 10000000)
```
The monthly payment would be:

```
Payment -32163.9520447
```

What just happened?

We calculated the monthly payment for a loan of 1 million at an annual rate of 10 percent. Given that we have 30 years to repay the loan, the `pmt` function tells us that we need to pay 32163.9520447 per month.

Number of payments

The NumPy `nper` function tells us how many periodic payments are necessary to pay off a loan. The required parameters are the interest rate of the loan, the fixed amount periodic payment, and the present value.

Time for action – determining the number of periodic payments

Consider a loan of 9000 at a rate of 10 percent with fixed monthly payments of 100:

1. **Getting the number of payments**: Find out how many payments are required with the NumPy `nper` function:

   ```
   print "Number of payments", numpy.nper(0.10/12, -100, 9000)
   ```

   The number of payments would be:
   ```
   Number of payments 167.047511801
   ```

What just happened?

We determined the number of payments needed to pay off a loan of 9000 with an interest rate of 10 percent and monthly payments of 100. The number of payments returned was 167.

Interest rate

The NumPy `rate` function calculates the interest rate given the number of periodic payments, the payment amount or amounts, the present value, and future value.

Time for action – figuring out the rate

Let's take the values from the previous *Time for action* tutorial and reverse compute the interest rate from the other parameters:

1. **Determining the rate**: Fill in the numbers from the previous *Time for action* tutorial:

   ```
   print "Interest rate", 12 * numpy.rate(167, -100, 9000, 0)
   ```

   The interest rate is approximately 10 percent as expected:
   ```
   Interest rate 0.0999756420664
   ```

What just happened?

We used the NumPy `rate` function and the values from the previous *Time for action* tutorial to compute the interest rate of the loan. Ignoring the rounding errors, we got the initial 10 percent we started with.

Window functions

Window functions are mathematical functions commonly used in signal processing. These functions are defined to be 0 outside a specified domain. NumPy has a number of window functions: `bartlett`, `blackman`, `hamming`, `hanning`, and `kaiser`. An example of the `hanning` function can be found in *Chapter 4, Convenience Functions for Your Convenience*.

Time for action – plotting the Bartlett window

The Bartlett window is a triangular smoothing window:

1. **Calculating the Bartlett window**: Call the NumPy `bartlett` function:

   ```
   window = numpy.bartlett(42)
   ```

2. **Plotting the Bartlett window**: Plotting is easy with Matplotlib:

   ```
   plot(window)
   show()
   ```

Here is the Bartlett window, which is triangular, as expected:

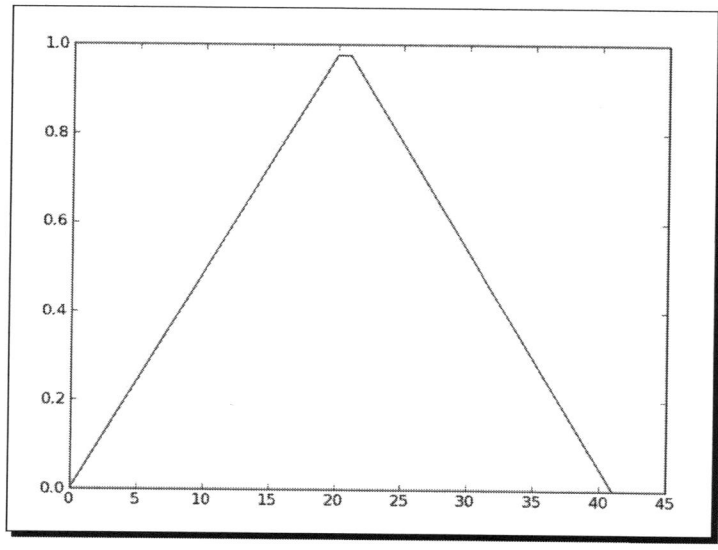

What just happened?

We plotted the Bartlett window with the NumPy `bartlett` function.

Blackman window

The Blackman window is formed by summing the first three terms of cosines:

$$w(n) = 0.42 - 0.5\cos(2\pi n/M) + 0.08\cos(4\pi n/M)$$

The NumPy `blackman` function returns the Blackman window. The only parameter is the number of points in the output window. If this number is 0 or less than 0, an empty array is returned.

Time for action – smoothing stock prices with the Blackman window

Let's smooth the close prices from the small AAPL stock prices datafile:

1. **Smoothing with the Blackman window**: Load the data into a NumPy array. Call the NumPy `blackman` function to form a window and then use this window to smooth the price signal:

```
closes=numpy.loadtxt('AAPL.csv', delimiter=',', usecols=(6,),
converters={1:datestr2num}, unpack=True)
N = int(sys.argv[1])
window = numpy.blackman(N)
smoothed = numpy.convolve(window/window.sum(),
   closes, mode='same')
```

2. **Plotting the Blackman window**: Plot the smoothed prices with Matplotlib. We will omit in this example the first five data points and the last five data points. The reason for this is that there is a strong boundary effect:

```
plot(smoothed[N:-N], lw=2, label="smoothed")
plot(closes[N:-N], label="closes")
legend(loc='best')
show()
```

The closing prices of AAPL smoothed with the Blackman window should appear as follows:

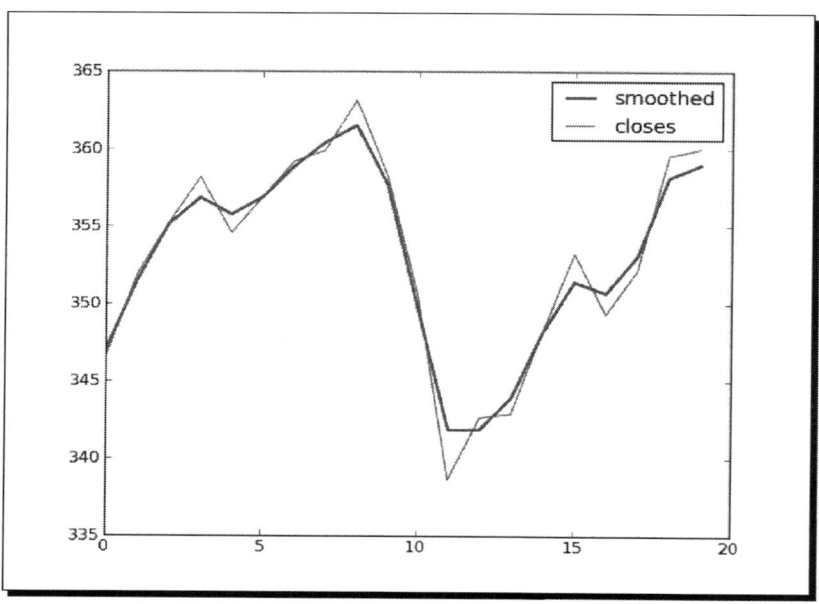

What just happened?

We plotted the closing price of AAPL from our sample data file that was smoothed using the Blackman window with the NumPy `blackman` function.

Hamming window

The Hamming window is formed by a weighted cosine. The formula is as follows:

$$w(n)=0.54+0.46cos\left(\frac{2\pi n}{M\text{-}1}\right) \quad 0 \le n \le M\text{-}1$$

The NumPy `hamming` function returns the Hamming window. The only parameter is the number of points in the output window. If this number is 0 or less than 0, an empty array is returned.

Time for action – plotting the Hamming window

Let's plot the Hamming window:

1. **Calculating the Hamming window**: Call the NumPy `hamming` function.

```
window = numpy.hamming(42)
```

2. **Plotting the Hamming window**: Plot the window with Matplotlib.

```
plot(window)
show()
```

The Hamming window plot is as follows:

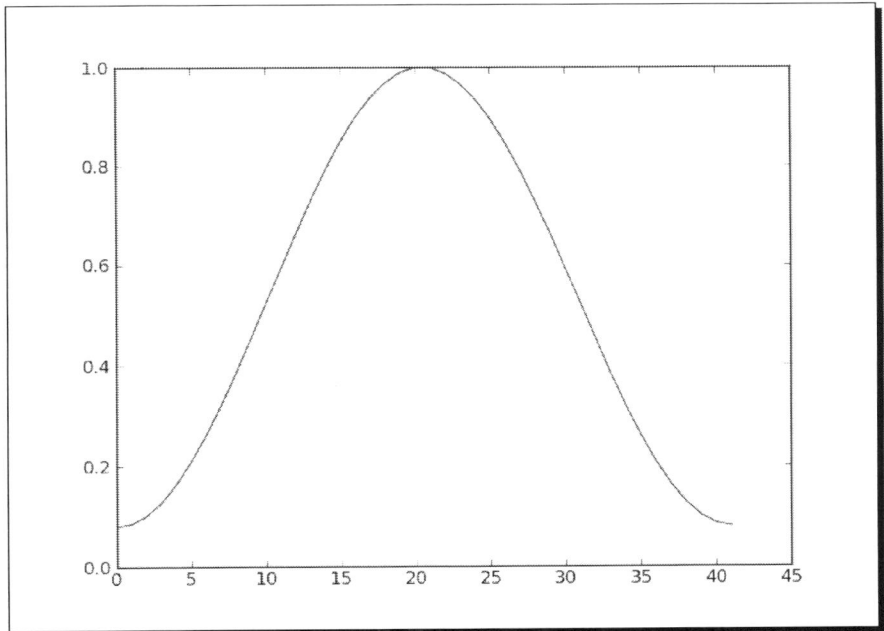

What just happened?

We plotted the Hamming window with the NumPy `hamming` function

Kaiser window

The Kaiser window is formed by the Bessel function. The formula is as follows:

$$w(n) = I_0\left(\beta \sqrt{1-\frac{4n^2}{(M-1)^2}}\right) / I_0(\beta)$$

Here I_0 is the zero order Bessel function The NumPy `kaiser` function returns the Kaiser window. The first parameter is the number of points in the output window. If this number is 0 or less than 0, an empty array is returned. The second parameter is the beta.

Time for action – plotting the Kaiser window

Let's plot the Kaiser window:

1. **Calculating the Kaiser window**: Call the NumPy `kaiser` function.

   ```
   window = numpy.kaiser(42, 14)
   ```

2. **Plotting the Kaiser window**: Plot the window with Matplotlib.

   ```
   plot(window)
   show()
   ```

The Kaiser window would appear as follows:

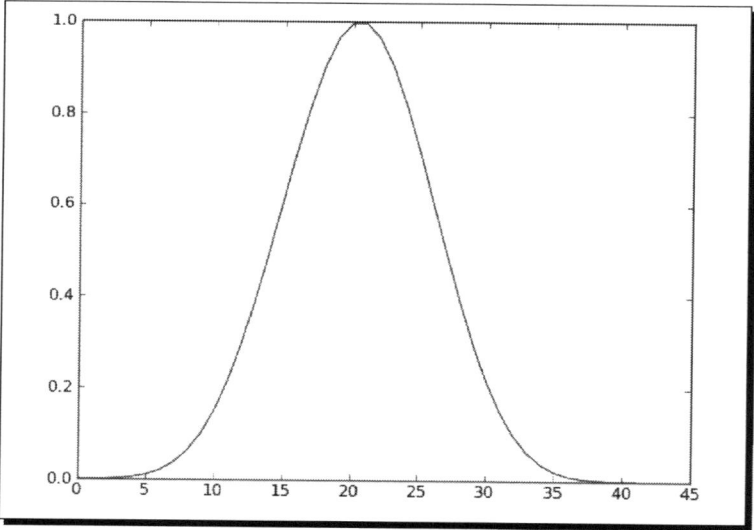

What just happened?

We plotted the Hamming window with the NumPy `kaiser` function.

Special mathematical functions

We will end this chapter with some special mathematical functions. First, the modified Bessel function of the first kind 0th order is represented in NumPy by `i0`. Second, the sinc function is represented in NumPy by a function with the same name.

Time for action – plotting the modified Bessel function

Let's see what the modified Bessel function of the first kind 0th order looks like:

1. **Calculate the x values**: Compute evenly spaced values with the NumPy `linspace` function.

    ```
    x = numpy.linspace(0, 4, 100)
    ```

2. **Calculate the function values**: Call the NumPy `i0` function.

    ```
    vals = numpy.i0(x)
    ```

3. **Plot the function**: Plot the modified Bessel function with Matplotlib:

    ```
    plot(x, vals)
    show()
    ```

The modified Bessel function would have the following output:

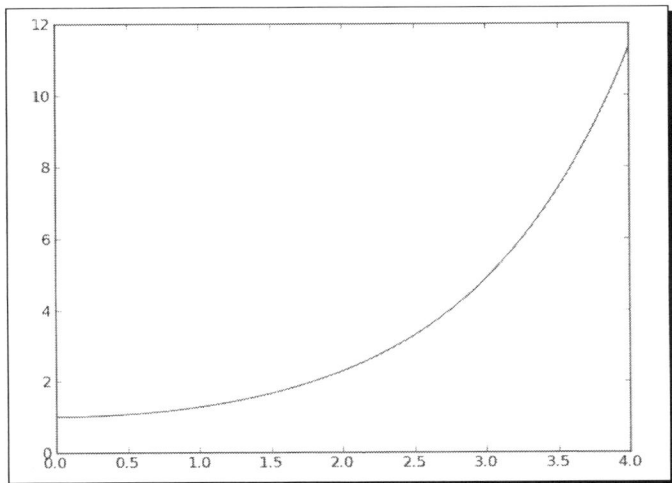

What just happened?

We plotted the modified Bessel function of the first kind 0th order with the NumPy `i0` function.

Sinc

The `sinc` function is widely used in mathematics and signal processing. NumPy has a function with the same name.

Time for action - plotting the sinc function

We will plot the `sinc` function:

1. **Compute the x values**: Compute evenly spaced values with the NumPy `linspace` function.

   ```
   x = numpy.linspace(0, 4, 100)
   ```

2. **Compute the function values**: Call the NumPy `sinc` function.

   ```
   vals = numpy.sinc(x)
   ```

3. **Plot the function**: Plot the `sinc` function with Matplotlib.

   ```
   plot(x, vals)
   show()
   ```

The `sinc` function would have the following output:

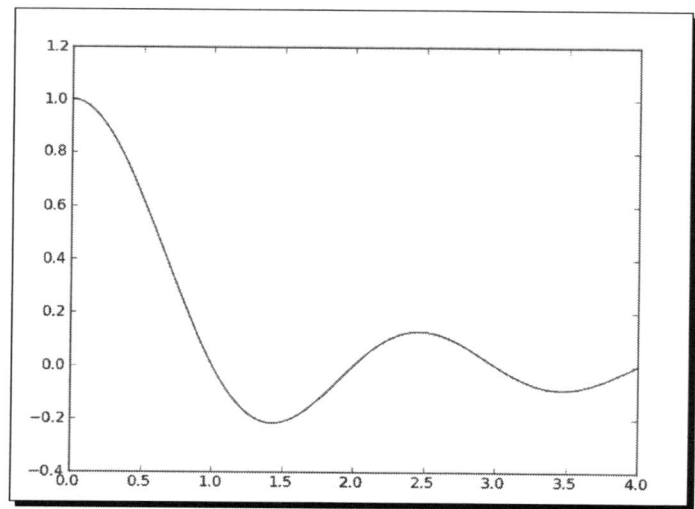

What just happened?

We plotted the well known `sinc` function with the NumPy `sinc` function.

Summary

This was a special chapter covering some of the more special NumPy topics. We covered sorting and searching, special functions, financial utilities, and window functions.

The next chapter will be about the very important subject of testing.

8
Assure Quality with Testing

Some programmers test only in production. If you are not one of them you're probably familiar with the concept of unit testing. Unit tests are automated tests written by a programmer to test his or her code. These tests could, for example, test a function or part of a function in isolation. Only a small unit of code is tested by each test. The benefits are increased confidence in the quality of the code, reproducible tests, and as a side effect, more clear code.

Python has good support for unit testing. Additionally, NumPy adds the `numpy.testing` *package to that for NumPy code unit testing.*

This chapter's topics include:

◆ Unit testing
◆ Asserts
◆ Floating point precision

Assert functions

The NumPy `testing` package has a number of utility functions that test whether a precondition is true or not:

Function	Description
assert_almost_equal	Raises an exception if two numbers are not equal up to a specified precision
assert_approx_equal	Raises an exception if two numbers are not equal up to a certain significance

Function	Description
assert_array_almost_equal	Raises an exception if two arrays are not equal up to a specified precision
assert_array_equal	Raises an exception if two arrays are not equal
assert_array_less	Raises an exception if two arrays do not have the same shape and the elements of the first array are strictly less than the elements of the second array
assert_equal	Raises an exception if two objects are not equal
assert_raises	Fails if a specified exception is not raised by a callable invoked with defined arguments
assert_warns	Fails if a specified warning is not thrown
assert_string_equal	Asserts that two strings are equal

Time for action – asserting almost equal

Imagine that you have two numbers that are almost equal. Let's use the `assert_almost_equal` function to check whether they are equal:

1. Call the function with low precision (up to 7 decimal places):

```
print "Decimal 6", numpy.testing.assert_almost_equal(0.123456789,
0.123456780, decimal=7)
```

Note that no exception is raised, as you can see in the following result:

```
Decimal 6 None
```

2. Call the function with high precision (up to 8 decimal places):

```
print "Decimal 7", numpy.testing.assert_almost_equal(0.123456789,
0.123456780, decimal=8)
```

The result is:

```
Decimal 7
Traceback (most recent call last):
    ...
    raise AssertionError(msg)
AssertionError:
Arrays are not almost equal
 ACTUAL: 0.123456789
 DESIRED: 0.12345678
```

What just happened?

We used the `assert_almost_equal` function from the NumPy `testing` package to check whether 0.123456789 and 0.123456780 are equal for different decimal precision.

1. Which parameter of the `assert_almost_equal` function specifies the decimal precision?

    a. `decimal`

    b. `precision`

    c. `tolerance`

    d. `significant`

Approximately equal arrays

The `assert_approx_equal` function raises an exception if two numbers are not equal up to a certain number of significant digits. The function result is an exception that is triggered by the condition:

```
abs(actual - expected) >= 10**-(significant - 1)
```

Time for action – asserting approximately equal

Let's take the numbers from the previous *Time for action* tutorial and let the `assert_approx_equal` function work on them:

1. Call the function with low significance:

```
print "Significance 8", numpy.testing.assert_approx_
equal(0.123456789, 0.123456780,
significant=8)
```

The result is:

Significance 8 None

2. Call the function with high significance:

```
print "Significance 9",
  numpy.testing.assert_approx_equal
  (0.123456789, 0.123456780, significant=9)
```

An exception is thrown:

```
Significance 9
Traceback (most recent call last):
  ...
    raise AssertionError(msg)
AssertionError:
Items are not equal to 9 significant digits:
 ACTUAL: 0.123456789
 DESIRED: 0.12345678
```

What just happened?

We used the `assert_approx_equal` function from the NumPy `testing` package to check whether 0.123456789 and 0.123456780 are equal for different decimal precision.

Almost equal arrays

The `assert_array_almost_equal` function raises an exception if two arrays are not equal up to a specified precision. The function checks whether the two arrays have the same shape. Then, the values of the arrays are compared element-by-element with:

$$|\text{expected - actual}| < 0.5 \ 10^{-\text{decimal}}$$

Time for action – asserting arrays almost equal

Let's form arrays with the values from the previous *Time for action* tutorial by adding a 0 to each array:

1. Calling the function with lower precision:

   ```
   print "Decimal 8", numpy.testing.assert_array_almost_equal([0,
     0.123456789], [0, 0.123456780], decimal=8)
   ```

   The result is:

   ```
   Decimal 8 None
   ```

2. Calling the function with higher precision:

   ```
   print "Decimal 9", numpy.testing.assert_array_almost_equal([0,
     0.123456789], [0, 0.123456780], decimal=9)
   ```

   An exception is thrown:

   ```
   Decimal 9
   Traceback (most recent call last):
     ...
   ```

```
assert_array_compare
    raise AssertionError(msg)
AssertionError:
Arrays are not almost equal


(mismatch 50.0%)
 x: array([ 0.        ,  0.12345679])
 y: array([ 0.        ,  0.12345678])
```

What just happened?

We compared two arrays with the NumPy `array_almost_equal` function.

Have a go hero – comparing array with different shapes

Use the NumPy `array_almost_equal` function to compare two arrays with different shapes.

Equal arrays

The `assert_array_equal` function raises an exception if two arrays are not equal. The shape of the arrays must have to be equal and the elements of each array must be equal. NaNs are allowed in the arrays. Alternatively, arrays can be compared with the `array_allclose` function. This function has the parameters `atol` (absolute tolerance) and `rtol` (relative tolerance). For two arrays a and b, these parameters satisfy the equation:

```
|a - b| <= (atol + rtol * |b|)
```

Time for action – comparing arrays

Let's compare two arrays with the functions we just mentioned. We will reuse the arrays from the previous *Time for action* tutorial and add a NaN to them:

1. Call the `array_allclose` function:
    ```
    print "Pass", numpy.testing.assert_allclose([0, 0.123456789,
      numpy.nan], [0, 0.123456780, numpy.nan], rtol=1e-7, atol=0)
    ```

    The result is:

    ```
    Pass None
    ```

2. Call the `array_equal` function:

```
print "Fail", numpy.testing.assert_array_equal([0, 0.123456789,
    numpy.nan], [0, 0.123456780, numpy.nan])
```

An exception is thrown:

```
Fail
Traceback (most recent call last):
    ...
assert_array_compare
    raise AssertionError(msg)
AssertionError:
Arrays are not equal

(mismatch 50.0%)
 x: array([ 0.        ,  0.12345679,          nan])
 y: array([ 0.        ,  0.12345678,          nan])
```

What just happened?

We compared two arrays with the `array_allclose` function and the `array_equal` function.

Ordering arrays

The `assert_array_less` function raises an exception if two arrays do not have the same shape and the elements of the first array are strictly less than the elements of the second array.

Time for action – checking the array order

Let's check whether one array is strictly greater than another array:

1. Call the `assert_array_less` function with two strictly ordered arrays

```
print "Pass", numpy.testing.assert_array_less([0, 0.123456789,
    numpy.nan], [1, 0.23456780, numpy.nan])
```

The result:

```
Pass None
```

2. Failing test: Call the `assert_array_less` function:

```
print "Fail", numpy.testing.assert_array_less([0, 0.123456789,
    numpy.nan], [0, 0.123456780, numpy.nan])
```

An exception is thrown:

```
Fail
Traceback (most recent call last):
  ...
    raise AssertionError(msg)
AssertionError:
Arrays are not less-ordered

(mismatch 100.0%)
 x: array([ 0.        ,  0.12345679,          nan])
 y: array([ 0.        ,  0.12345678,          nan])
```

What just happened?

We checked the ordering of two arrays with the `assert_array_less` function.

Objects comparison

The `assert_equal` function raises an exception if two objects are not equal. The objects do not have to be NumPy arrays, they can also be lists, tuples, or dictionaries.

Time for action – comparing objects

Suppose you need to compare two tuples. We can use the `assert_equal` function to do that:

1. Call the `assert_equal` function:

    ```
    print "Equal?", numpy.testing.assert_equal((1, 2), (1, 3))
    ```

    An exception is thrown:

    ```
    Equal?
    Traceback (most recent call last):
      ...
        raise AssertionError(msg)
    AssertionError:
    Items are not equal:
    item=1

     ACTUAL: 2
     DESIRED: 3
    ```

What just happened?

We compared two tuples with the `assert_equal` function—an exception was raised because the tuples were not equal to each other.

String comparison

The `assert_string_equal` function asserts that two strings are equal. If the test fails an exception is thrown and the difference between the strings is shown. The case of the string characters matters.

Time for action – comparing strings

Let's compare strings. Both strings are the word "NumPy":

1. Call the `assert_string_equal` function to compare a string with itself. This test, of course, should pass:

   ```
   print "Pass", numpy.testing.assert_string_equal("NumPy", "NumPy")
   ```

   The test passes:

   Pass None

2. Call the `assert_string_equal` function to compare a string with another string with the same letters but different casing. This test should throw an exception:

   ```
   print "Fail", numpy.testing.assert_string_equal("NumPy", "Numpy")
   ```

   An exception is thrown:

   ```
   Fail

   Traceback (most recent call last):

       ...

       raise AssertionError(msg)

   AssertionError: Differences in strings:

   - NumPy?        ^

   + Numpy?        ^
   ```

What just happened?

We compared two strings with the `assert_string_equal` function. The test threw an exception when the casing did not match.

Floating point comparisons

The `assert_array_almost_equal_nulp` and `assert_array_max_ulp` NumPy functions provide consistent floating point comparisons. **ULP** stands for **Unit of Least Precision** of floating point numbers. According to the IEEE 754 specification, a half ULP precision is required for elementary arithmetic operations.

Machine epsilon is the largest relative rounding error in floating point arithmetic. Machine epsilon is equal to ULP relative to 1. The NumPy `finfo` function allows us to determine the machine epsilon.

Time for action – comparing with assert_array_almost_equal_nulp

Let's see the `assert_array_almost_equal_nulp` function in action:

1. Determine the machine epsilon with the `finfo` function:

```
eps = numpy.finfo(float).eps
print "EPS", eps
```

The epsilon would be:

```
EPS 2.22044604925e-16
```

2. Compare two almost equal floats: Compare 1.0 with 1 + epsilon using the `assert_almost_equal_nulp` function. Do the same for 1 + 2 * epsilon:

```
print "1",
    numpy.testing.assert_array_almost_equal_nulp(1.0, 1.0 + eps)
print "2",
    numpy.testing.assert_array_almost_equal_nulp(1.0, 1.0 + 2 * eps)
```

The result:

```
1 None

2

Traceback (most recent call last):

  ...

 assert_array_almost_equal_nulp

    raise AssertionError(msg)

AssertionError: X and Y are not equal to 1 ULP (max is 2)
```

What just happened?

We determined the machine epsilon with the `finfo` function. We then compared 1.0 with
1 + epsilon with the `assert_almost_equal_nulp` function. This test passed, however
adding a little bit more resulted in an exception.

Comparison of floats with more ULPs

The `assert_array_max_ulp` function allows you to specify an upper bound for the
number ULPs you would allow. The `maxulp` parameter accepts an integer value for the limit.
The value of this parameter is 1 by default.

Time for action – comparing using maxulp of 2

Let's do the same comparisons as in the previous *Time for action* tutorial, but specify a
`maxulp` of 2 when necessary:

1. Determine the machine epsilon with the `finfo` function:

```
eps = numpy.finfo(float).eps
print "EPS", eps
```

The epsilon would be:

```
EPS 2.22044604925e-16
```

2. Do the comparisons as done in the previous *Time for action* tutorial, but use the
`assert_array_max_ulp` function with the appropriate `maxulp` value:

```
print "1", numpy.testing.assert_array_max_ulp(1.0, 1.0 + eps)
print "2", numpy.testing.assert_array_max_ulp(1.0, 1 + 2 * eps,
  maxulp=2)
```

The output:

```
1 1.0
2 2.0
```

What just happened?

We compared the same values as the previous *Time for action* tutorial, but specified a
`maxulp` of 2 in the second comparison. Using the `assert_array_max_ulp` function with
the appropriate `maxulp` value, these tests passed with a return value of the number of ULPs.

Summary

We learned about testing and NumPy testing utilities in this chapter. We covered unit testing, assert functions and floating point precision.

The topic of the next chapter is Matplotlib—the Python scientific visualization and graphing library.

9
Plotting with Matplotlib

Matplotlib is a very useful python plotting library. It integrates nicely with NumPy but is a separate open source project. You can find a gallery of beautiful examples at `http://matplotlib.sourceforge.net/gallery.html`.

Matplotlib also has utility functions to download and manipulate data from Yahoo Finance. We will see several examples of stock charts.

This chapter features extended coverage of:

- ◆ Simple plots
- ◆ Subplots
- ◆ Histograms
- ◆ Plot Customization
- ◆ Logplots

Simple plots

The `matplotlib.pyplot` package contains functionality for simple plots. It is important to remember that each subsequent function call changes the state of the current plot. Eventually we will want to either save the plot in a file or display it with the `show` function.

Time for action – plotting a polynomial function

To illustrate how plotting works, let's display some polynomial graphs. We will use the NumPy polynomial function `poly1d` to create a polynomial.

1. **Create the polynomial**: Take the standard input values as polynomial coefficients. Use the NumPy `poly1d` function to create a polynomial.

   ```
   func = numpy.poly1d(numpy.array(sys.argv[1:]).astype(float))
   ```

2. **Create the x values**: Create the x values with the NumPy `linspace` function. Use the range -10 to 10 and create 30 even spaced values.

   ```
   x = numpy.linspace(-10, 10, 30)
   ```

3. **Calculate the polynomial values**: Calculate the polynomial values using the polynomial that we created in the first step.

   ```
   y = func(x)
   ```

4. **Call the plot function**: Call the `plot` function, this does not immediately display the graph.

   ```
   pyplot.plot(x, y)
   ```

5. **Add a label to the x axis**: Add a label to the x axis with `xlabel` function.

   ```
   pyplot.xlabel('x')
   ```

6. **Add a label to the y axis**: Add a label to the y axis with `ylabel` function.

   ```
   pyplot.ylabel('y(x)')
   ```

7. **Display the plot on the screen**: Call the `show` function to display the graph.

   ```
   pyplot.show()
   ```

Here is a plot with polynomial coefficients 1, 2, 3, and 4:

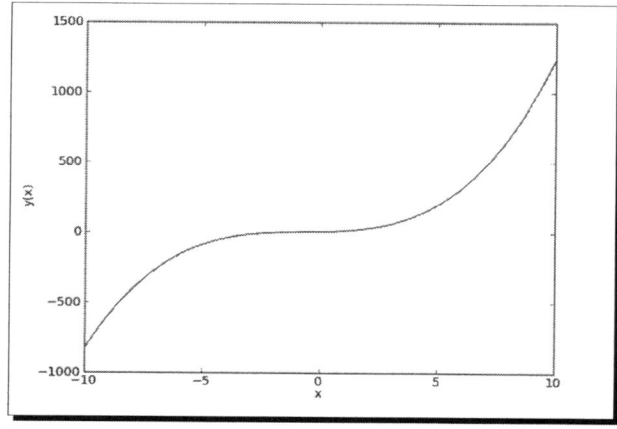

What just happened?

We displayed a graph of a polynomial on our screen. We added labels to the x and y axis.

Pop quiz – doing the thing

1. What does the `plot` function do?

    a. It displays two-dimensional plots on screen

    b. It saves an image of a two-dimensional plot in a file

    c. It does both a and b...

    d. It does neither a, b, or c

Plot format string

The `plot` function accepts an unlimited number of arguments. In the previous section we gave it two arrays as arguments. We could also specify the line color and style with an optional format string. By default it is a solid blue line denoted as `b-`, but you can specify a different color and style such as red dashes.

Time for action – plotting a polynomial and its derivative

Let's plot a polynomial and its first order derivative using the `derive` function with m as 1. We already did the first part in the previous *Time for action* tutorial. We want to have two different line styles to be able to discern what is what.

1. **Differentiate**: Create and differentiate the polynomial.

```
func = numpy.poly1d(numpy.array(sys.argv[1:]).astype(float))
func1 = func.deriv(m=1)
x = numpy.linspace(-10, 10, 30)
y = func(x)
y1 = func1(x)
```

2. **Plot the polynomial and its derivative**: Plot the polynomial and its derivative in two different styles: red circles and green dashes. You cannot see the colors in a print copy of this book so you will have to try it out for yourself.

```
pyplot.plot(x, y, 'ro', x, y1, 'g--')
pyplot.xlabel('x')
pyplot.ylabel('y')
pyplot.show()
```

The graph again with polynomial coefficients 1, 2, 3, and 4:

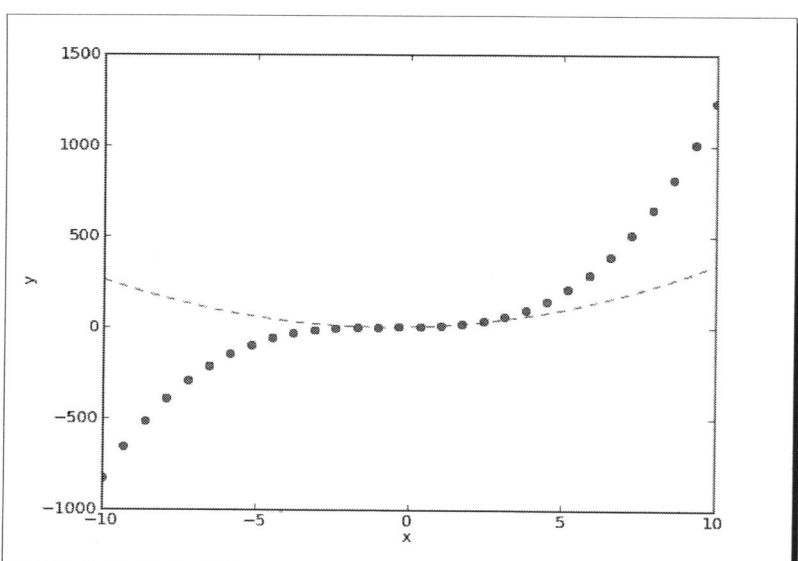

What just happened?

We plotted a polynomial and its derivative using two different line styles and one call of the `plot` function.

Subplots

At a certain point you will have too many lines in one plot. Still you would like to have everything grouped together. We can achieve this with the `subplot` function.

Time for action – plotting a polynomial and its derivatives

Let's plot a polynomial and its first and second derivative. We will make three subplots for the sake of clarity:

1. **Create the polynomial and its derivatives**: Create a polynomial and its derivatives using the following code.

```
func = numpy.poly1d(numpy.array(sys.argv[1:]).astype(float))
x = numpy.linspace(-10, 10, 30)
y = func(x)
func1 = func.deriv(m=1)
```

```
y1 = func1(x)
func2 = func.deriv(m=2)
y2 = func2(x)
```

2. **Create the first subplot**: Create the first subplot of the polynomial with the `subplot` function. The first parameter of this function is the number of rows, the second parameter is the number of columns, and the third parameter is an index number starting with 1. Alternatively, you can combine the three parameters into a single number such as 311. The subplots will be organized in 3 rows and 1 column. Give the subplot the title "Polynomial". Make a solid red line.

```
pyplot.subplot(311)
pyplot.plot(x, y, 'r-')
pyplot.title("Polynomial")
```

3. **Create the second subplot**: Create the third subplot of the first derivative with the `subplot` function. Give the subplot the title "First Derivative". Use a line of blue triangles.

```
pyplot.subplot(312)
pyplot.plot(x, y1, 'b^')
pyplot.title("First Derivative")
```

4. **Create the third subplot**: Create the second subplot of the second derivative with the `subplot` function. Give the subplot the title "Second Derivative". Use a line of green circles.

```
pyplot.subplot(313)
pyplot.plot(x, y2, 'go')
pyplot.title("Second Derivative")
pyplot.xlabel('x')
pyplot.ylabel('y')
pyplot.show()
```

The three subplots with polynomial coefficients 1, 2, 3, and 4:

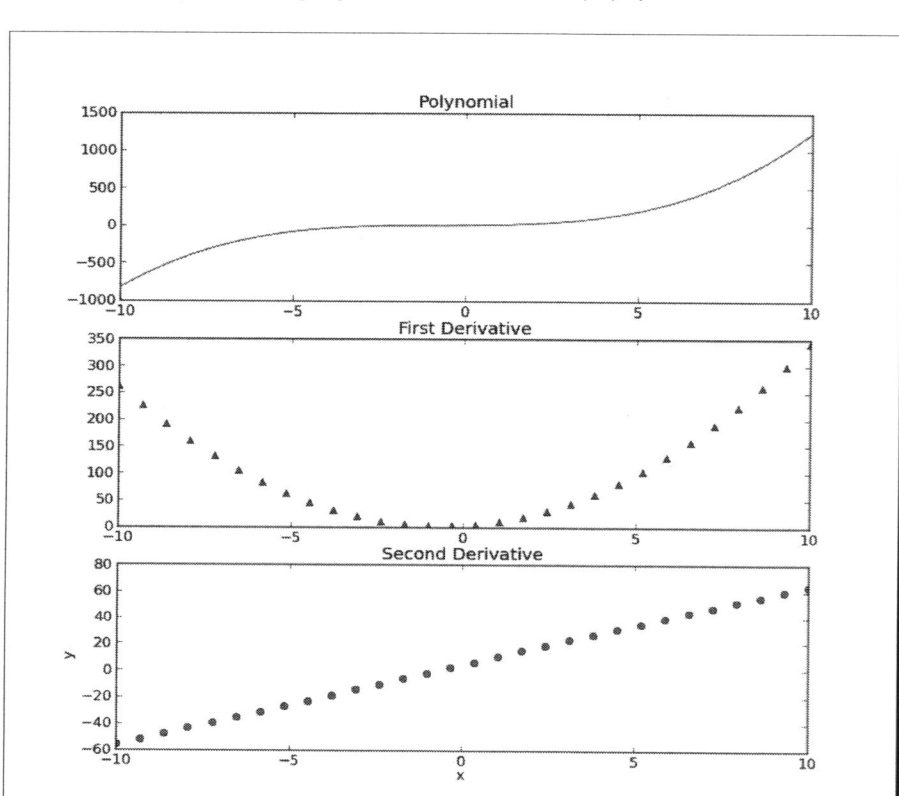

What just happened?

We plotted a polynomial and its first and second derivative using three different line styles and three subplots in 3 rows and 1 column.

Finance

Matplotlib can help us monitor our stock investments. The `matplotlib.finance` package has utilities with which we can download stock quotes from Yahoo Finance (http://finance.yahoo.com/). The data can then be plotted as candlesticks.

Time for action – plotting a year's worth of stock quotes

We can plot a year's worth of stock quotes data with the `matplotlib.finance` package. This will require a connection to Yahoo Finance, which will be the data source.

1. **Determine start date**: Determine the start date by subtracting 1 year from today.

```
today = date.today()
start = (today.year - 1, today.month, today.day)
```

2. **Create locators**: We need to create so-called locators. These objects from the `matplotlib.dates` package are needed to locate months and days on the x-axis.

```
alldays = DayLocator()
months = MonthLocator()
```

3. **Create a formatter**: Create a date formatter to format the dates on the x-axis. This formatter will create a string containing the short name of a month and the year.

```
month_formatter = DateFormatter("%b %Y")
```

4. **Download the quotes**: Download the stock quote data from Yahoo finance with the code below:

```
quotes = quotes_historical_yahoo(sys.argv[1], start, today)
```

5. **Create a figure**: Create a Matplotlib `figure` object—this is a top level container for plot components.

```
fig = pyplot.figure()
```

6. **Add a subplot**: Add a subplot to the figure.

```
ax = fig.add_subplot(111)
```

7. **Set the major locator**: Set the major locator on the x axis to the months locator. This locator is responsible for the big ticks on the x-axis.

```
ax.xaxis.set_major_locator(months)
```

8. **Set the minor locator**: Set the minor locator on the x axis to the days locator. This locator is responsible for the small ticks on the x-axis.

```
ax.xaxis.set_minor_locator(alldays)
```

9. **Set the major formatter**: Set the major formatter on the x axis to the months formatter. This formatter is responsible for the labels of the big ticks on the x axis.

```
ax.xaxis.set_major_formatter(month_formatter)
```

10. **Create the candlesticks**: A function in the `matplotlib.finance` package allows us to display candlesticks. Create the candlesticks using the quotes data. It is possible to specify the width of the candlesticks. For now use the default value.

```
candlestick(ax, quotes)
```

11. **Format the x axis labels as dates**: Format the labels on the x-axis as dates. This should rotate the labels on the x axis, so that they fit better.

```
fig.autofmt_xdate()
pyplot.show()
```

The candlestick chart for DISH (Dish Network Corp.) would appear as follows:

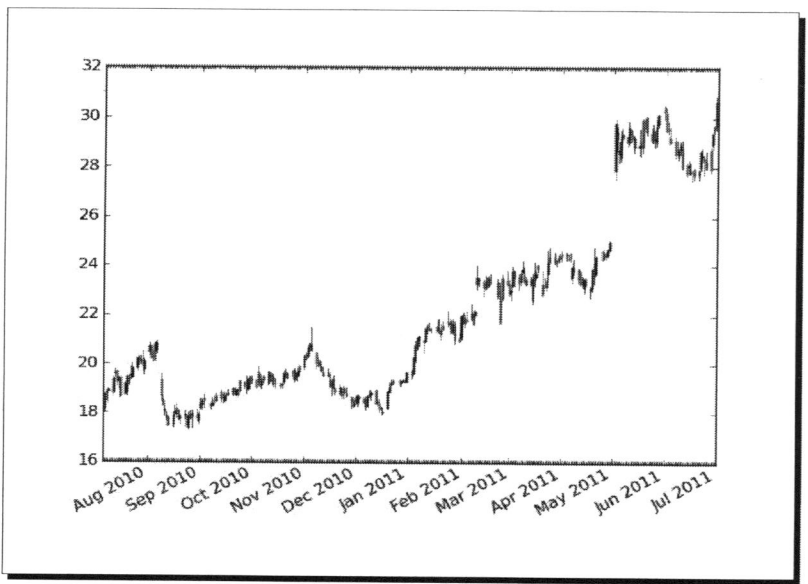

What just happened?

We downloaded a year's worth of data from Yahoo Finance. We charted this data using candlesticks.

Histograms

Histograms visualize the distribution of numerical data. Matplotlib has the handy `hist` function that graphs histograms. The `hist` function has two arguments—the array containing the data and the number of bars.

Time for action – charting stock price distributions

Let's chart the stock price distribution of quotes from Yahoo Finance.

1. **Download the data**: Download the data going back 1 year.

   ```
   today = date.today()
   start = (today.year - 1, today.month, today.day)

   quotes = quotes_historical_yahoo(sys.argv[1], start, today)
   ```

2. **Extract the close price**: The quotes data in the previous step is stored in a Python list. Convert this to a NumPy array and extract the close prices.

   ```
   quotes = numpy.array(quotes)
   close = quotes.T[4]
   ```

3. **Draw the histogram**: Draw the histogram with a reasonable number of bars.

   ```
   pyplot.hist(close, numpy.sqrt(len(close)))
   pyplot.show()
   ```

The histogram for DISH would appear as follows:

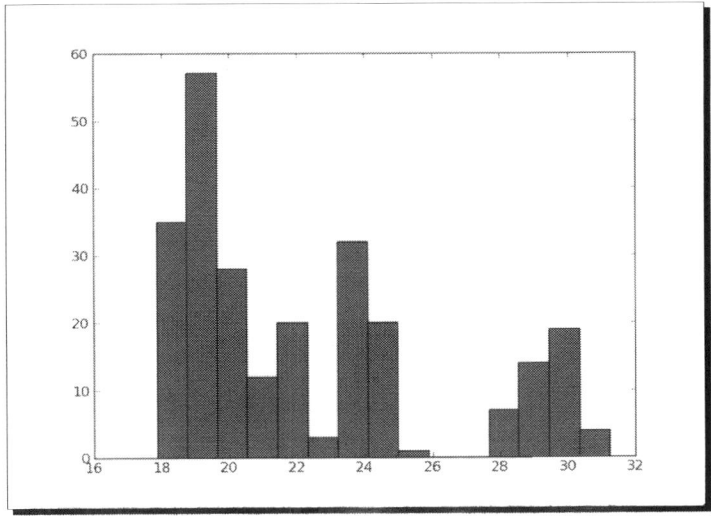

What just happened?

We charted the stock price distribution of DISH as histogram.

Have a go hero – drawing a bell curve

Overlay a bell curve using the average price and standard deviation. This is, of course, only an exercise.

Logarithmic plots

Logarithmic plots are useful when the data has a wide range of values. Matplotlib has the functions `semilogx` (logarithmic x axis), `semilogy` (logarithmic y axis), and `loglog` (x and y axis logarithmic).

Time for action – plotting stock volume

Stock volume varies a lot, so let's plot it on a logarithmic scale. First we need to download historical data from Yahoo Finance, extract the dates and volume, create locators and a date formatter, create the figure, and add it a subplot. We already went through these steps in the previous *Time for action* tutorial, so we will skip them here.

1. **Logarithmic plot**: Plot the volume using a logarithmic scale.

   ```
   pyplot.semilogy(dates, volume)
   ```

   Now set the locators and format the x-axis as dates. Instructions for these steps can be found in the previous *Time for action* tutorial as well. The stock volume using a logarithmic scale for DISH would appear as follows:

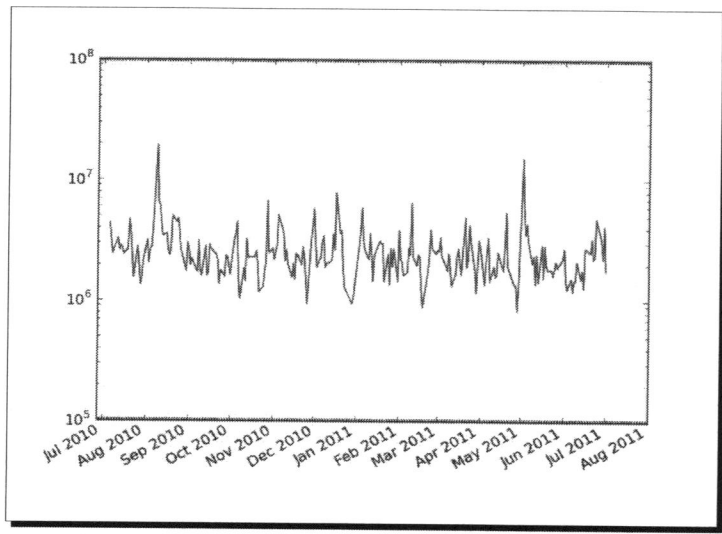

What just happened?

We plotted stock volume using a logarithmic scale.

Scatter plots

A scatter plot displays values for two numerical variables in the same data set. The Matplotlib `scatter` function creates a scatter plot. Optionally: we can specify color and size of the data points in the plot as well as alpha transparency.

Time for action – plotting price and volume returns with scatter plot

We can easily make a scatter plot of the stock price and volume returns. Again let's download the necessary data from Yahoo Finance.

1. **Extract the close price and volume**: The quotes data in the previous step is stored in a Python list. Convert this to a NumPy array and extract the close and volume values.

```
dates = quotes.T[4]
volume = quotes.T[5]
```

2. **Calculate the returns**: Calculate the close price and volume returns.

```
ret = numpy.diff(close)/close[:-1]
volchange = numpy.diff(volume)/volume[:-1]
```

3. **Create a figure**: Create a Matplotlib figure object.

```
fig = pyplot.figure()
```

4. **Add a subplot**: Add a subplot to the figure.

```
ax = fig.add_subplot(111)
```

5. **Create the scatter plot**: Create the scatter plot with the color of the data points linked to the close return, and the size linked to the volume change.

```
ax.scatter(ret, volchange, c=ret * 100,
   s=volchange * 100, alpha=0.5)
```

6. **Title and grid**: Set the title of the plot and put a grid on it.

```
ax.set_title('Close and volume returns')
ax.grid(True)

pyplot.show()
```

The scatter plot for DISH will appear as follows:

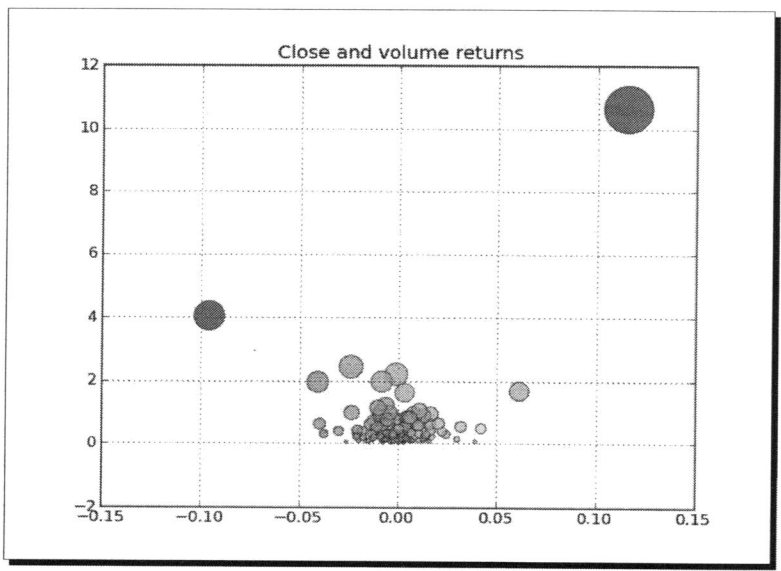

What just happened?

We made a scatter plot of the close price and volume returns for DISH.

Fill between

The `fill_between` function fills a region of a plot with a specified color. We can also choose an alpha channel value. The function also has a `where` parameter so that we can shade a region based on a condition.

Time for action – shading plot regions based on a condition

Imagine that you want to shade the region of a stock chart, where the closing price is below average with different color than when it is above the mean. The `fill_between` function is the best choice for the job. We will again omit the steps of downloading historical data going back 1 year, extracting dates and close prices, creating locators and date formatter.

1. **Create a figure**: Create a Matplotlib figure object.

   ```
   fig = pyplot.figure()
   ```

2. **Add a subplot**: Add a subplot to the figure.

   ```
   ax = fig.add_subplot(111)
   ```

3. **Plot the closing price**: Plot the closing price.

   ```
   ax.plot(dates, close)
   ```

4. **Call fill_between**: Shade the regions of the plot below the closing price using different colors depending whether the values are below or above the average price.

   ```
   pyplot.fill_between(dates, close.min(), close,
       where=close>close.mean(), facecolor="green", alpha=0.4)
   pyplot.fill_between(dates, close.min(), close,
       where=close<close.mean(), facecolor="red", alpha=0.4)
   ```

Now we can finish the plot by setting locators and formatting the x-axis values as dates. The stock price using conditional shading for DISH:

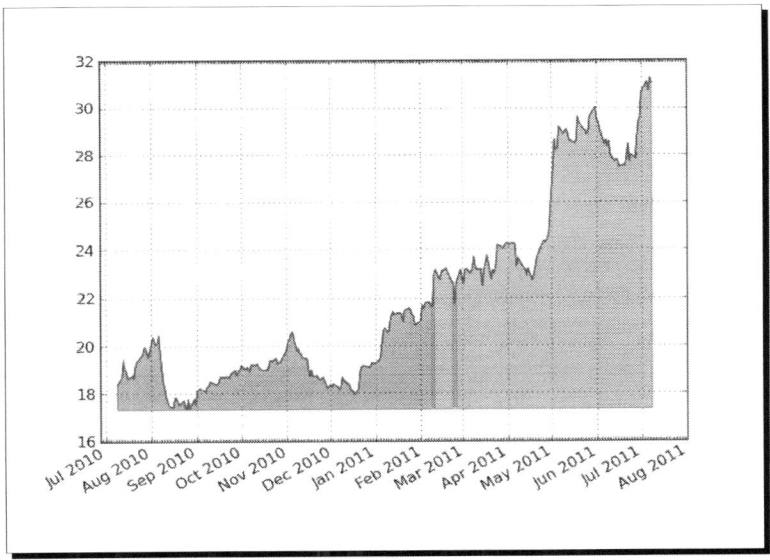

What just happened?

We shaded the region of a stock chart, where the closing price is below average with different color than when it is above the mean.

Legend and annotations

Legends and annotations are essential for good plots. We can create transparent legends with the `legend` function and let Matplotlib figure out where to place them. Also with the `annotate` function we can put annotations very accurately on a plot. There are a large number of annotation and arrow styles.

Time for action – using legend and annotations

In Chapter 3, *Get Into Terms with Commonly Used Functions* we learned how to calculate the exponential moving average of stock prices. We will plot the close price of a stock and three of its exponential moving averages. To clarify the plot, we will add a legend. Also, we will indicate crossovers of two of the averages with annotations. Some steps are again omitted to avoid repetition.

1. **Calculate and plot the exponential moving averages**: Go back to *Chapter 3, Get into Terms with Commonly Used Functions* if needed and review the exponential moving average algorithm. Calculate and plot the exponential moving averages of 9, 12 and 15 periods.

```python
emas = []
for i in range(9, 18, 3):
    weights = numpy.exp(numpy.linspace(-1., 0., i))
    weights /= weights.sum()

    ema = numpy.convolve(weights, close)[i-1:-i+1]
    idx = (i - 6)/3
    ax.plot(dates[i-1:], ema, lw=idx, label="EMA(%s)" % (i))
    data = numpy.column_stack((dates[i-1:], ema))
    emas.append(numpy.rec.fromrecords(
       data, names=["dates", "ema"]))
```

   Notice that the `plot` function call needs a label for the legend. We stored the moving averages in record arrays for the next step.

2. **Find the crossover points**: Let's find the crossover points of the first two moving averages.

```python
first = emas[0]["ema"].flatten()
second = emas[1]["ema"].flatten()
bools = numpy.abs(first[-len(second):] - second)/second < 0.0001
xpoints = numpy.compress(bools, emas[1])
```

3. **Annotate the crossover points**: Now that we have the crossover points annotate them with arrows. Make sure that the annotation text is slightly away from the crossover points.

```
for xpoint in xpoints:
    ax.annotate('x', xy=xpoint, textcoords='offset points',
                xytext=(-50, 30),
                arrowprops=dict(arrowstyle="->"))
```

4. **Add a legend**: Add a legend and let Matplotlib decide where to put it.

```
leg = ax.legend(loc='best', fancybox=True)
```

5. **Make the legend transparent**: Make the legend transparent by setting the alpha channel value.

```
leg.get_frame().set_alpha(0.5)
```

The stock price and moving averages with legend and annotations would appear as follows:

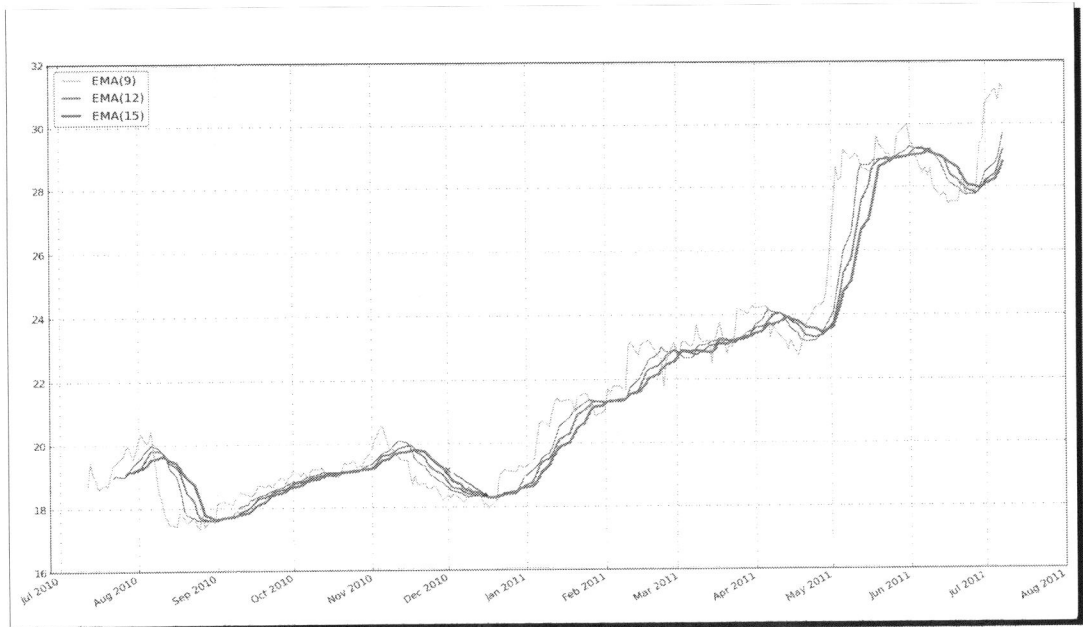

What just happened?

We plotted the close price of a stock and three of its exponential moving averages. We added a legend to the plot. We annotated the crossover points of the first two averages with annotations.

Summary

This chapter was about Matplotlib—a Python plotting library. We covered simple plots, histograms, plot customization, subplots and logplots. We also saw a few examples of displaying stock charts.

The next chapter is about SciPy—a scientific Python framework that is built on top of NumPy.

10
When NumPy is Not Enough: SciPy and Beyond

SciPy is built on top of NumPy. It adds functionality such as numerical integration, optimization, statistics, and special functions.

In this chapter we will cover the following topics:

- ◆ File I/O
- ◆ Statistics
- ◆ Signal processing
- ◆ Optimization
- ◆ Interpolation
- ◆ Image processing

Matlab and Octave

Matlab and its open source alternative Octave are popular mathematical programs. The `scipy.io` package has functions that let you load Matlab or Octave code in Python programs and vice versa. The `loadmat` function loads a `.mat` file. The `savemat` function saves a dictionary of names and arrays into a `.mat` file.

Time for action – saving and loading a .mat file

If we start with NumPy code and decide to use the said code within a Matlab or Octave environment, the easiest thing to do is create a `.mat` file. We then can load the file within Matlab or Octave. Let's go through the necessary steps.

1. **Call savemat**: Create a NumPy array and call `savemat` to create a `.mat` file. This function has two parameters: a file name, and a dictionary containing variable names and values.

    ```
    a = numpy.arange(7)

    scipy.io.savemat("a.mat", {"array": a})
    ```

2. **Load the .mat file**: Within a Matlab or Octave environment, load the `.mat` file and check the stored array.

    ```
    octave-3.4.0:7> load a.mat
    octave-3.4.0:8> a

    octave-3.4.0:8> array
    array =

       0
       1
       2
       3
       4
       5
       6
    ```

What just happened?

We created a `.mat` file from NumPy code and loaded it within Octave. We checked the NumPy array that was created.

Pop quiz – loading .mat files

1. Which function loads *.mat* files?

    a. `Loadmatlab`

    b. `loadmat`

    c. `loadoct`

    d. `frommat`

Statistics

The SciPy statistics module is called `scipy.stats`. There is one class that implements continuous distributions and one class that implements discrete distributions. Also in this module, functions can be found that can perform a great number of statistical tests.

Time for action – analyzing random values

We will generate random values that mimic a normal distribution and analyze the generated data with statistical functions from the `scipy.stats` package.

1. **Generate random values**: Generate random values from a normal distribution using the `scipy.stats` package.

```
generated = scipy.stats.norm.rvs(size=900)
```

2. **Fit the values**: Fit the generated values to a normal distribution. This basically gives us the mean and standard deviation of the data set.

```
print "Mean", "Std", scipy.stats.norm.fit(generated)
```

The mean and standard deviation would be:

Mean Std (0.0071293257063200707, 0.95537708218972528)

3. **Skewness test**: Perform a skewness test. This test returns two values. The second value is the p value—the probability that the skewness of the data set corresponds to a normal distribution. P values range from 0 to 1.

```
print "Skewtest", "pvalue", scipy.stats.skewtest(generated)
```

The result of the skewness test would be:

Skewtest pvalue (-0.62120640688766893, 0.5344638245033837)

So there is a 53 percent chance that we are dealing with a normal distribution.

4. **Kurtosis test**: Perform a kurtosis test. This test is setup similarly to the skewness test, but of course, applies to kurtosis.

```
print "Kurtosistest", "pvalue",
  scipy.stats.kurtosistest(generated)
```

The result of the kurtosis test would be:

Kurtosistest pvalue (1.3065381019536981, 0.19136963054975586)

5. **Normality test**: Perform a normality test. This test also returns two values, of which the second is a p value.

```
print "Normaltest", "pvalue", scipy.stats.normaltest(generated)
```

The result of the normality test would be:

Normaltest pvalue (2.09293921181506, 0.35117535059841687)

6. **Score at percentile**: We can find the value at a certain percentile easily with SciPy.

```
print "95 percentile",
   scipy.stats.scoreatpercentile(generated, 95)
```

The value at the 95th percentile would be:

95 percentile 1.54048860252

7. **Percentile of score**: Do the opposite of the previous step to find the percentile at 1.

```
print "Percentile at 1",
   scipy.stats.percentileofscore(generated, 1)
```

The percentile at 1 would be:

Percentile at 1 85.5555555556

8. **Plot with Matplotlib**: Plot the generated values in a histogram with Matplotlib. More information about Matplotlib can be found in the previous chapter.

```
matplotlib.pyplot.hist(generated)
matplotlib.pyplot.show()
```

The histogram of the generated random values is as follows:

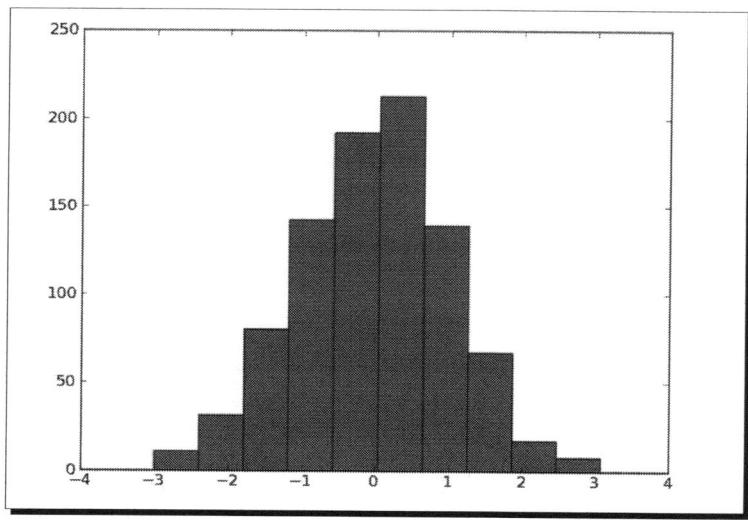

What just happened?

We created a data set from a normal distribution and analyzed it with the `scipy.stats` module.

Have a go hero – improving the data generation

Judging from the histogram in the previous *Time for action* tutorial, there is still room for improvement when it comes to generating the data. Try using NumPy or different parameters of the `scipy.stats.norm.rvs` function.

Samples comparison and SciKits

Often we will have two data samples, maybe from different experiments, that are somehow related. Statistical tests exist that can compare the samples. Some of these have been implemented in the `scipy.stats` module.

Another statistical test that I like is the *Jarque Bera* normality test from `scikits.statsmodels.stattools`. **SciKits** are small experimental Python software toolkits. **They are not part of SciPy**.

Time for action – comparing stock log returns

We will download the stock quotes for the last year of two trackers using Matplotlib. As mentioned in the previous chapter we can retrieve quotes from Yahoo Finance. We will compare the log returns of the close price of DIA and SPY. Also we will perform the Jarque Bera test on the difference of the log returns.

1. **Download quotes**: Write a function that can return the close price for a specified stock.

```
def get_close(symbol):
    today = date.today()
    start = (today.year - 1, today.month, today.day)

    quotes = quotes_historical_yahoo(symbol, start, today)
    quotes = numpy.array(quotes)

    return quotes.T[4]
```

2. **Calculate log returns**: Calculate the log returns for DIA and SPY. The log returns are calculated by taking the natural logarithm of the close price and then taking the difference of consecutive values.

```
spy =   numpy.diff(numpy.log(get_close("SPY")))
dia =   numpy.diff(numpy.log(get_close("DIA")))
```

3. **Compare means**: The means comparison test checks whether two different samples could have the same mean value. Two values are returned, of which the second is the p value from 0 to 1.

```
print "Means comparison", scipy.stats.ttest_ind(spy, dia)
```

The result of the means comparison test would be:

```
Means comparison (-0.017995865641886155, 0.98564930169871368)
```

So there is about a 98 percent chance that the two samples have the same mean log return.

4. **Kolmogorov Smirnov test**: The *Kolmogorov Smirnov* two samples test tells us how likely it is that two samples are drawn from the same distribution.

```
print "Kolmogorov smirnov test", scipy.stats.ks_2samp(spy, dia)
```

Again two values are returned of which the second value is the p value.

```
Kolmogorov smirnov test (0.063492063492063516,
0.67615647616238039)
```

5. **Jarque Bera test**: Unleash the Jarque Bera normality test on the difference of the log returns.

```
print "Jarque Bera test",
   scikits.statsmodels.stattools.jarque_bera(spy - dia)[1]
```

The p value of the Jarque Bera normality test would be:

```
Jarque Bera test 0.596125711042
```

6. **Plot histograms with Matplotlib**: Plot the histograms of the log returns and the difference thereof with Matplotlib.

```
matplotlib.pyplot.hist(spy, histtype="step", lw=1, label="SPY")
matplotlib.pyplot.hist(dia, histtype="step", lw=2, label="DIA")
matplotlib.pyplot.hist(spy - dia, histtype="step", lw=3,
   label="Delta")
matplotlib.pyplot.legend()
matplotlib.pyplot.show()
```

The histograms of the log returns and difference is as follows:

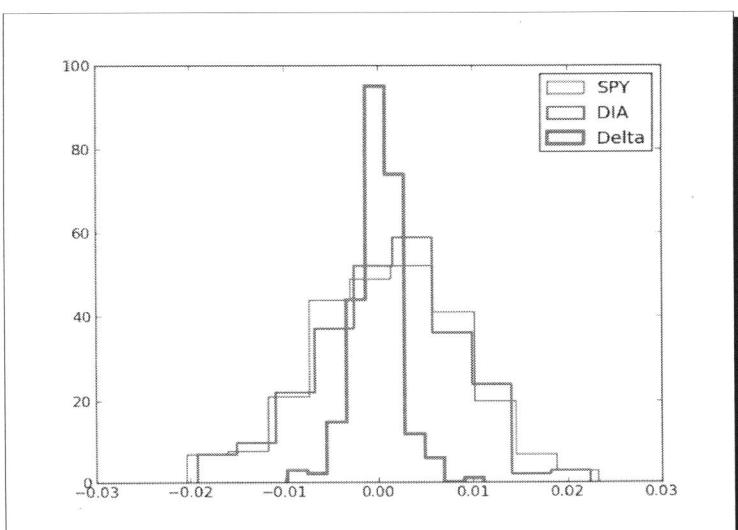

What just happened?

We compared samples of log returns for DIA and SPY. Also we performed the Jarque Bera test on the difference of the log returns.

Signal processing

The `scipy.signal` module contains filter functions and B-spline interpolation algorithms. A SciPy signal is defined as an array of numbers. An example of a filter is the `detrend` function. This function takes a signal and does a linear fit on it. This trend is then subtracted from the original input data.

Time for action – detecting a trend in QQQ

Often we are more interested in the trend of a data sample than in detrending it. Still we can get the trend back easily after detrending. Let's do that for 1 year of price data for QQQ:

1. **Download quotes**: Write code that gets the close price and corresponding dates for QQQ.

```
today = date.today()
start = (today.year - 1, today.month, today.day)

quotes = quotes_historical_yahoo("QQQ", start, today)
```

```
quotes = numpy.array(quotes)
dates = quotes.T[0]
qqq = quotes.T[4]
```

2. **Detrend the signal**: Detrend the signal.

```
y = scipy.signal.detrend(qqq)
```

3. **Create locators**: Create month and day locators for the dates.

```
alldays = DayLocator()
months = MonthLocator()
```

4. **Date formatter**: Create a date formatter that creates a string of month name and year.

```
month_formatter = DateFormatter("%b %Y")
```

5. **Figure and subplot**: Create a figure and subplot.

```
fig = matplotlib.pyplot.figure()
ax = fig.add_subplot(111)
```

6. **Data and underlying trend**: Plot the data and underlying trend by subtracting the detrended signal.

```
matplotlib.pyplot.plot(dates, qqq, 'o', dates, qqq - y, '-')
```

7. **Locators and formatter**: Set the locators and formatter.

```
ax.xaxis.set_minor_locator(alldays)
ax.xaxis.set_major_locator(months)
ax.xaxis.set_major_formatter(month_formatter)
```

8. **X axis labels**: Format the x axis labels as dates.

```
fig.autofmt_xdate()
matplotlib.pyplot.show()
```

The following figure shows the QQQ prices with a trend line:

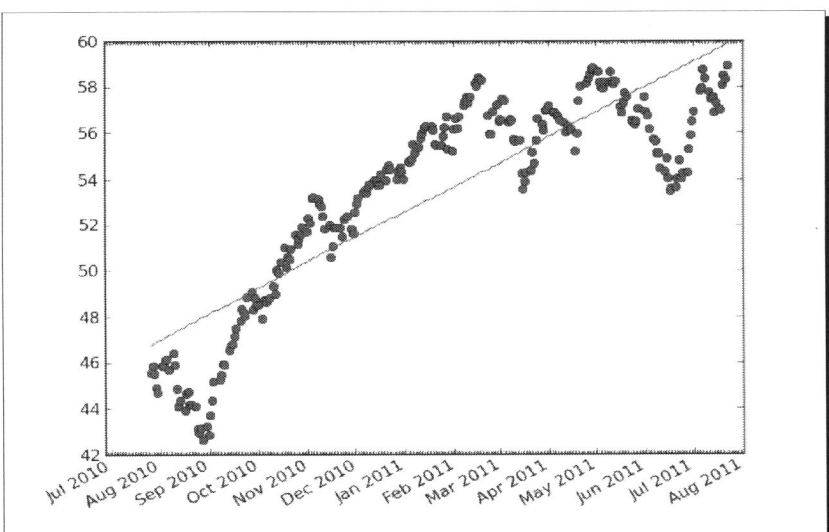

What just happened?

We plotted the closing price for QQQ with a trend line.

Fourier analysis

Signals in the real world often have a periodic nature. A commonly used tool to deal with these signals is the Fourier transform. Functions for Fourier transforms can be found in the `scipy.fftpack` module. Included in the package are fast Fourier transforms, differential and pseudo-differential operators, as well as several helper functions. Matlab users will be pleased to know that a number of functions in the `scipy.fftpack` module have the same name as their Matlab counterparts and similar function as their Matlab equivalents.

Time for action – filtering a detrended signal

We learned in the previous *Time for action* tutorial how to detrend a signal. This detrended signal could have a cyclical component. Let's try to visualize this. Some of the steps are a repetition of steps in the previous *Time for action* tutorial, such as downloading the data and setting up Matplotlib objects. These steps are omitted here.

1. **Frequency spectrum**: Apply Fourier transforms, giving us the frequency spectrum.

```
amps = numpy.abs(scipy.fftpack.fftshift(scipy.fftpack.rfft(y)))
```

2. **Noise filter**: Filter out the noise. Let's say if the magnitude of a frequency component is below 10 percent of the strongest component, throw it out:

```
amps[amps < 0.1 * amps.max()] = 0
```

3. **Inverse transform**: Transform the filtered signal back to the original domain and plot it together with the detrended signal.

```
matplotlib.pyplot.plot(dates, y, 'o', label="detrended")
matplotlib.pyplot.plot(dates,
  -scipy.fftpack.irfft(scipy.fftpack.ifftshift(amps)),
  label="filtered")
```

4. **X axis labels**: Format the x axis labels as dates and add a legend:

```
fig.autofmt_xdate()
matplotlib.pyplot.legend()
```

5. **Second subplot**: Add a second subplot and plot the frequency spectrum after filtering.

```
ax2 = fig.add_subplot(212)
N = len(qqq)
matplotlib.pyplot.plot(numpy.linspace(-N/2, N/2, N), amps,
  label="transformed")
```

6. **Legend**: Display the legend and plot.

```
matplotlib.pyplot.legend()
matplotlib.pyplot.show()
```

The following plots are of the signal and frequency spectrum:

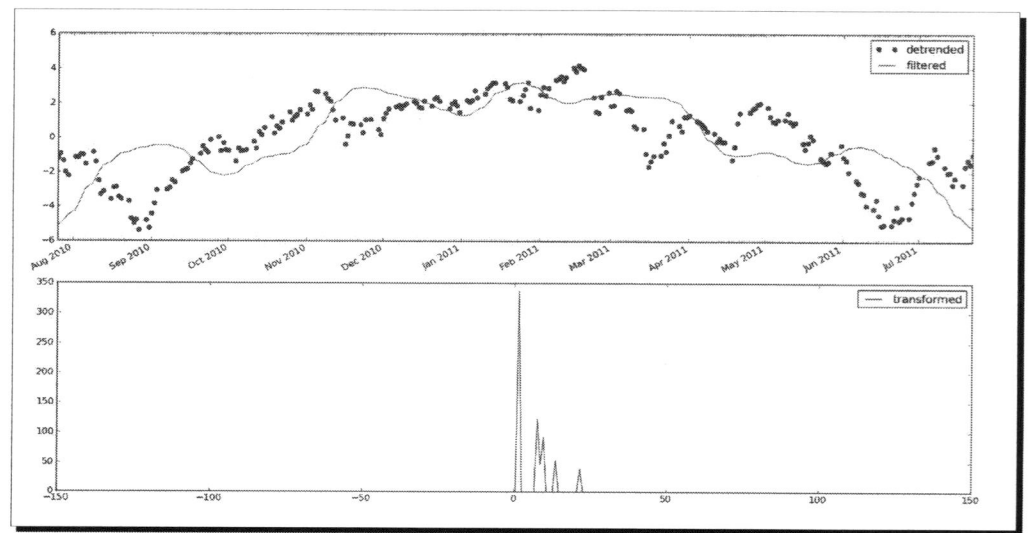

What just happened?

We detrended a signal and applied a simple filter on it using the `scipy.fftpack` module.

Optimization

Several optimization algorithms are provided by the `scipy.optimize` module. One of the algorithms is a least squares fitting function `leastsq`. When calling this function, we are required to provide a residuals function. This function is used to minimize the sum of the squares of the residuals. Also, it is necessary to give the algorithm a starting point. This should be a best guess—as close as possible to the real solution. Otherwise execution will stop after about 800 iterations.

Time for action – fitting to a sine

In the previous *Time for action* tutorial we created a simple filter for detrended data. Now let's use a more restrictive filter that will leave us only with the main frequency component. We will fit a sinusoidal pattern to it and plot our results. This model has four parameters—amplitude, frequency, phase, and vertical offset.

1. **Download quotes**: Write code that gets the close price and corresponding dates for QQQ.

    ```
    today = date.today()
    start = (today.year - 1, today.month, today.day)

    quotes = quotes_historical_yahoo("QQQ", start, today)
    quotes = numpy.array(quotes)

    dates = quotes.T[0]
    qqq = quotes.T[4]
    ```

2. **Detrended signal**: Detrend the signal.

    ```
    y = scipy.signal.detrend(qqq)
    ```

3. **Locators**: Create month and day locators for the dates.

    ```
    alldays = DayLocator()
    months = MonthLocator()
    ```

4. **Date formatter**: Create a date formatter that creates a string of month name and year.

    ```
    month_formatter = DateFormatter("%b %Y")
    ```

5. **Figure and subplot**: Create a figure and subplot.

```
fig = matplotlib.pyplot.figure()
ax = fig.add_subplot(211)
```

6. **Locators and formatter**: Set the locators and formatter.

```
ax.xaxis.set_minor_locator(alldays)
ax.xaxis.set_major_locator(months)
ax.xaxis.set_major_formatter(month_formatter)
```

7. **Frequency spectrum**: Apply Fourier transforms giving us the frequency spectrum.

```
amps = numpy.abs(scipy.fftpack.fftshiftn(scipy.fftpack.rfft(y)))
```

8. **Main component**: Retrieve the main component of the frequency spectrum.

```
amps[amps < amps.max()] = 0
```

9. **Residual functions**: Define a residuals function based on a sine wave model.

```
def residuals(p, y, x):
    A,k,theta,b = p
    err = y-A * numpy.sin(2* numpy.pi* k * x + theta) + b
    return err
```

10. **Inverse transform**: Transform the filtered signal back to the original domain.

```
filtered = -scipy.fftpack.irfft(scipy.fftpack.ifftshift(amps))
```

11. **Initial guess**: Guess the values of the parameters we are trying to estimate.

```
N = len(qqq)
f = numpy.linspace(-N/2, N/2, N)
p0 = [filtered.max(), f[amps.argmax()]/(2*N), 0, 0]
print "P0", p0
```

The initial values would be:

```
P0 [2.6679532410065212, 0.00099598469163686377, 0, 0]
```

12. **Least squares fit**: Call the `leastsq` function.

```
plsq = scipy.optimize.leastsq(residuals, p0, args=(filtered,
    dates))
p = plsq[0]
print "P", p
```

The final parameter values are:

```
P [  2.67678014e+00   2.73033206e-03  -8.00007036e+03
 -5.01260321e-03]
```

13. **First subplot**: Finish the first subplot with detrended data, filtered data, and fit of the filtered data. Use a date format for the horizontal axis and add a legend.

```
matplotlib.pyplot.plot(dates, y, 'o', label="detrended")
matplotlib.pyplot.plot(dates, filtered, label="filtered")
matplotlib.pyplot.plot(dates, p[0] * numpy.sin(2 * numpy.pi *
   dates * p[1] + p[2]) + p[3], '^', label="fit")
fig.autofmt_xdate()
matplotlib.pyplot.legend()
```

14. **Second subplot**: Add a second subplot with a legend of the main component of the frequency spectrum.

```
ax2 = fig.add_subplot(212)
matplotlib.pyplot.plot(f, amps, label="transformed")

matplotlib.pyplot.legend()
matplotlib.pyplot.show()
```

The following are the resulting charts:

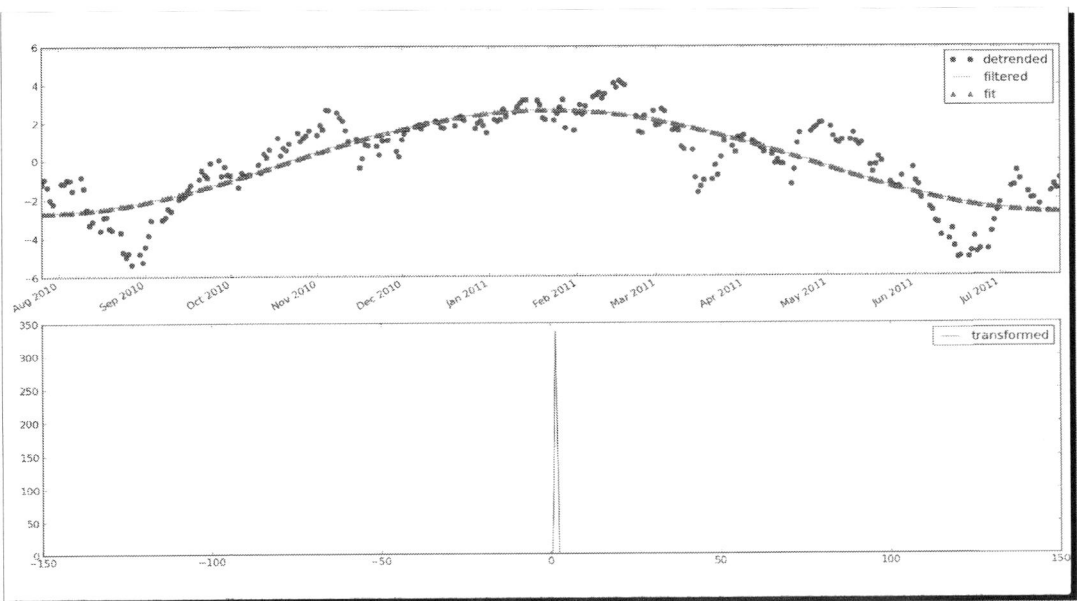

What just happened?

We detrended 1 year of price data for QQQ. This signal was then filtered until only the main component of the frequency spectrum was left over. We fitted a sine to the filtered signal using the `scipy.optimize` module.

Numerical integration

SciPy has a numerical integration package `scipy.integrate`, that has no equivalent in NumPy. The `quad` function can integrate a one variable function between two points. These points can be at infinity.

Time for action – calculating the Gaussian integral

The Gaussian integral is related to the `error` function, but has no finite limits. It evaluates to the square root of pi. Let's calculate the integral with the `quad` function.

1. **Quad function**: Calculate the Gaussian integral with the `quad` function.

```
print "Gaussian integral", numpy.sqrt(numpy.pi),
scipy.integrate.quad(lambda x: numpy.exp(-x**2),
-numpy.inf, numpy.inf)
```

The return value is the outcome and its error would be:

```
Gaussian integral 1.77245385091 (1.7724538509055159,
1.4202636780944923e-08)
```

What just happened?

We calculated the Gaussian integral with the `quad` function.

Interpolation

The `scipy.interpolate` function interpolates a function based on experimental data. The `interp1d` class can create a linear or cubic interpolation function. By default a linear interpolation function is constructed, but if the `kind` parameter is set, a cubic interpolation function is created instead. The `interp2d` class works the same way, but in 2D.

Time for action – interpolating in one dimension

We will create data points using a `sinc` function and add some random noise to it. After that, we will do a linear and cubic interpolation, and plot the results.

1. **Data points**: Create the data points and add noise to it.

```
x = numpy.linspace(-18, 18, 36)
noise = 0.1 * numpy.random.random(len(x))
signal = numpy.sinc(x) + noise
```

2. **Linear interpolation**: Create a linear interpolation function and apply it to an input array with five times as many data points.

```
interpreted = scipy.interpolate.interp1d(x, signal)
x2 = numpy.linspace(-18, 18, 180)
y = interpreted(x2)
```

3. **Cubic interpolation**: Do the same as in the previous step, but with cubic interpolation.

```
cubic = scipy.interpolate.interp1d(x, signal, kind="cubic")
y2 = cubic(x2)
```

4. **Plot**: Plot the results with Matplotlib.

```
matplotlib.pyplot.plot(x, signal, 'o', label="data")
matplotlib.pyplot.plot(x2, y, '-', label="linear")
matplotlib.pyplot.plot(x2, y2, '-', lw=2, label="cubic")

matplotlib.pyplot.legend()
matplotlib.pyplot.show()
```

The following diagram is a plot of the data, linear, and cubic interpolation:

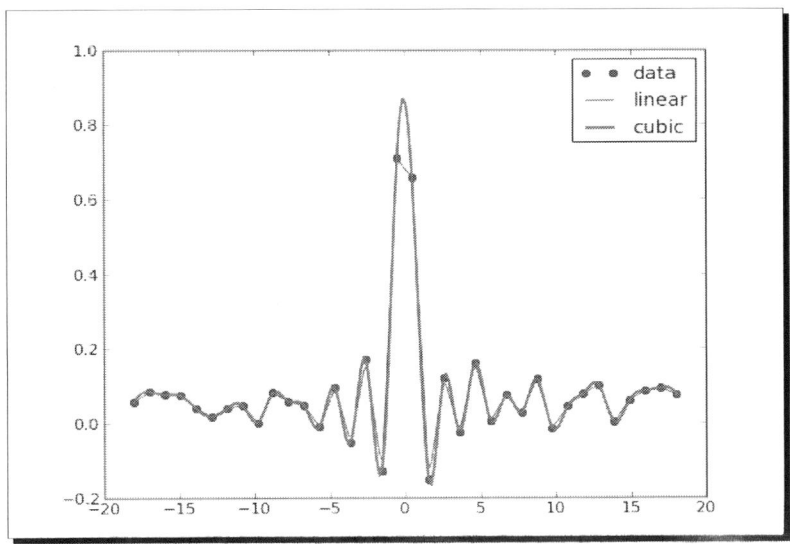

What just happened?

We created a data set from the `sinc` function and added noise to it. We then did linear and cubic interpolation using the `interp1d` class of the `scipy.interpolate` module.

Image processing

With SciPy, we can do image processing using the `scipy.ndimage` package. The module contains various image filters and utilities.

Time for action – manipulating Lena

In the `scipy.misc` module, there is a utility which loads the image of "Lena". We will apply some filters on this image and rotate it.

1. **Lena**: Load the "Lena" image and display it in a subplot.

```
image = scipy.misc.lena().astype(numpy.float32)

matplotlib.pyplot.subplot(221)
matplotlib.pyplot.title("Original Image")
img = matplotlib.pyplot.imshow(image)
```

Note that we are dealing with a `float32` array.

2. **Median filter**: Apply a median filter to the image and display it in a second subplot.

```
matplotlib.pyplot.subplot(222)
matplotlib.pyplot.title("Median Filter")
filtered = scipy.ndimage.median_filter(image, size=(42,42))
matplotlib.pyplot.imshow(filtered)
```

3. **Rotation**: Rotate the image and display it in the third subplot.

```
matplotlib.pyplot.subplot(223)
matplotlib.pyplot.title("Rotated")
rotated = scipy.ndimage.rotate(image, 90)
matplotlib.pyplot.imshow(rotated)
```

4. **Prewitt filter**: Apply a Prewitt filter to the image and display it in the fourth subplot.

```
matplotlib.pyplot.subplot(224)
matplotlib.pyplot.title("Prewitt Filter")
filtered = scipy.ndimage.prewitt(image)
matplotlib.pyplot.imshow(filtered)
matplotlib.pyplot.show()
```

The following are the resulting images:

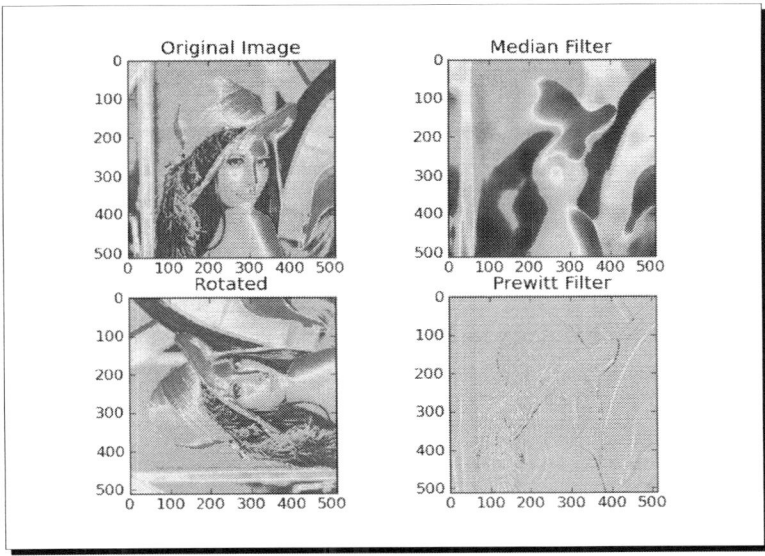

What just happened?

We manipulated the image of "Lena" in several ways using the `scipy.ndimage` module.

Summary

In this chapter we only scratched the surface of what is possible with SciPy and SciKits. Still, we learned a bit about file I/O, statistics, signal processing, optimization, interpolation, and image processing.

Pop Quiz Answers

Chapter 1, NumPy Quick Start

What does `arrange(5)` do?	It creates a NumPy array with values 0 to 4.
	The created NumPy array has values 0, 1, 2, 3, 4.

Chapter 2, Beginning with NumPy Fundamentals

How is the shape of an `ndarray` stored?	It is stored in a tuple.

Chapter 3, Get into Terms with Commonly Used Functions

Which function returns the weighted average of an array?	`average`

Chapter 4, Convenience Functions for Your Convenience

Which function returns the covariance of two arrays?	`cov`

Chapter 5, Working with Matrices and ufuncs

What is the row delimiter in a string accepted by the `mat` and `bmat` functions?	Semicolon

Chapter 6, Move Further with NumPy Modules

Which function can create matrices?	`mat`

Chapter 7, Peeking into Special Routines

Which NumPy module deals with random numbers?	`random`

Chapter 8, Assured Quality with Testing

Which parameter of the `assert_almost_equal` function specifies the decimal precision?	`decimal`

Chapter 9, Plotting with Matplotlib

What does the `plot` function do?	It does neither a, b, or c.

Chapter 10, When NumPy is not enough SciPy and Beyond

Which function loads `.mat` files?	`loadmat`

Index

correlation
  about 82
  computing, for stock returns 82-85
correlation coefficient 83
covariance 82
cov function 82
CSV files
  about 50
  data, loading from 51
cumprod function 79

D

data
  about 49
  fitting, to polynomial 85-87
  loading, from CSV files 51
  summarizing 61-64
data type 26
data type objects 30
dates
  about 58
  dealing with 58-61
datestr2num function 59
datetime object 59
Debian
  about 13
  NumPy, installing on 13
  Python, installing on 10
depth stacking 39
depth-wise splitting 42
determinant
  about 124
  calculating, of matrix 124
det function 124
detrend function 187
diag function 122
diagonal function 82, 85
diff function 57, 58, 88
DISH
  histogram 173
distribution (distro) 13
divide function 106
DMG file 14
dot function 74, 119, 120
dsplit function 41, 42
dstack function 39

dtype class
  about 32
  attributes 32
dtype constructors 31, 32

E

eigenvalues
  about 120
  calculating 120, 121
eigenvectors
  about 120
  calculating 120, 121
eig function 120
elements
  extracting, from array 139
  selecting, of array 28
ellipsis
  used, for slicing 35
equal universal function 115
error function 194
exp function 69
exponential moving average
  about 68
  calculating 69, 70
  switching to 72
extract function 138, 139
extremums 86
eye function 50

F

factorial
  about 79
  calculating 79
Fast Fourier transform
  about 124
  calculating 125, 126
features, IPython 20
Fedora 13
fft function 125, 126
fftshift function 126
Fibonacci matrix
  creating 108
Fibonacci numbers
  about 108
  computing 108, 109
  computing, with matrix 109

Thank you for buying
NumPy 1.5 Beginner's Guide

About Packt Publishing

Packt, pronounced 'packed', published its first book "*Mastering phpMyAdmin for Effective MySQL Management*" in April 2004 and subsequently continued to specialize in publishing highly focused books on specific technologies and solutions.

Our books and publications share the experiences of your fellow IT professionals in adapting and customizing today's systems, applications, and frameworks. Our solution based books give you the knowledge and power to customize the software and technologies you're using to get the job done. Packt books are more specific and less general than the IT books you have seen in the past. Our unique business model allows us to bring you more focused information, giving you more of what you need to know, and less of what you don't.

Packt is a modern, yet unique publishing company, which focuses on producing quality, cutting-edge books for communities of developers, administrators, and newbies alike. For more information, please visit our website: www.packtpub.com.

About Packt Open Source

In 2010, Packt launched two new brands, Packt Open Source and Packt Enterprise, in order to continue its focus on specialization. This book is part of the Packt Open Source brand, home to books published on software built around Open Source licences, and offering information to anybody from advanced developers to budding web designers. The Open Source brand also runs Packt's Open Source Royalty Scheme, by which Packt gives a royalty to each Open Source project about whose software a book is sold.

Writing for Packt

We welcome all inquiries from people who are interested in authoring. Book proposals should be sent to author@packtpub.com. If your book idea is still at an early stage and you would like to discuss it first before writing a formal book proposal, contact us; one of our commissioning editors will get in touch with you.

We're not just looking for published authors; if you have strong technical skills but no writing experience, our experienced editors can help you develop a writing career, or simply get some additional reward for your expertise.

Sage Beginner's Guide

ISBN: 978-1-84951-446-0   Paperback: 364 pages

Unlock the full potential of Sage for simplifying and automating mathematical computing

1. The best way to learn Sage which is a open source alternative to Magma, Maple, Mathematica, and Matlab

2. Learn to use symbolic and numerical computation to simplify your work and produce publication-quality graphics

3. Numerically solve systems of equations, find roots, and analyze data from experiments or simulations

Python Testing Cookbook

ISBN: 978-1-84951-466-8   Paperback: 364 pages

Over 70 simple but incredibly effective recipes for taking control of automated testing using powerful Python testing tools

1. Learn to write tests at every level using a variety of Python testing tools

2. The first book to include detailed screenshots and recipes for using Jenkins continuous integration server (formerly known as Hudson)

3. Explore innovative ways to introduce automated testing to legacy systems

4. Written by Greg L. Turnquist – senior software engineer and author of Spring Python 1.1

Please check **www.PacktPub.com** for information on our titles

Nginx 1 Web Server Implementation Cookbook

ISBN: 978-1-84951-496-5   Paperback: 336 pages

Over 100 recipes to master using the Nginx HTTP server and reverse proxy

1. Quick recipes and practical techniques to help you maximize your experience with Nginx

2. Interesting recipes that will help you optimize your web stack and get more out of your existing setup

3. Secure your website and prevent your setup from being compromised using SSL and rate-limiting techniques

4. Get more out of Nginx by using it as an important part of your web application using third-party modules

Python Geospatial Development

ISBN: 978-1-84951-154-4   Paperback: 508 pages

Build a complete and sophisticated mapping application from scratch using Python tools for GIS development

1. Build applications for GIS development using Python

2. Analyze and visualize Geo-Spatial data

3. Comprehensive coverage of key GIS concepts

4. Recommended best practices for storing spatial data in a database

5. Draw maps, place data points onto a map, and interact with maps

Please check **www.PacktPub.com** for information on our titles

Printed in Great Britain
by Amazon.co.uk, Ltd.,
Marston Gate.